For Hiram Drache,
valued colleague and
cherished friend, with best
wishes and warmest regards.
Dave Danbom
September 1990.

"Our Purpose is to Serve"

9-7-90
To A Real
Pioneer!
[illegible]

The First Century of the North Dakota Agricultural Experiment Station

David B. Danbom

North Dakota Institute for Regional Studies

NDSU

100

Printed in the United States of America

Library of Congress Catalog Card Number: 89-63044
ISBN 0-91042-38-5

The cover photo, the first two photo sections and the second photo in the third section are courtesy of the University Archives, NDSU.
All other photos are courtesy of the
Agricultural Communications Department, NDSU.

For Elizabeth and Mark,
Native North Dakotans

CONTENTS

INTRODUCTION

"Our Purpose is to Serve" is a history of the first century of operation of the North Dakota Agricultural Experiment Station. The NDAES is part of a national network of state research facilities in agriculture funded initially by Congress in the Hatch Act of 1887. In the period since the enactment of that keystone legislation, Congress has passed other laws that have affected the NDAES and its sister institutions in a variety of ways. The NDAES has also felt the effects of being part of a national—even international—scientific community in which standards are set, judgments are made, researchers are trained, and so forth. As a consequence, the story of the NDAES must necessarily also be the story of the larger scientific and political matrix of which it is a small part.

But the NDAES is also an integral part of the state of North Dakota and of the agriculture that has been at the heart of that state's existence for over a century. What makes the NDAES unique is not that it is part of a national scientific enterprise. It is unique because it is the only research institution in the world devoted to North Dakota agriculture and its problems. The unique nature of the NDAES dictates that its story will be intertwined with the story of North Dakota and its agricultural system.

For a century North Dakota has lived by its agriculture, and so, too, has the NDAES. When North Dakota agriculture has thrived the Station has prospered. When North Dakota agriculture has been depressed the Station has suffered. In general North Dakota is a poor state, and the NDAES has reflected that poverty in its continual struggles for adequate funding. This, too, is part of the Station's story.

The programs of the NDAES have always reflected the recognition by the makers of scientific policy of the centrality of agricultural prosperity to the material well being of the state. The Station has been devoted throughout its history to enhancing income and income security among the farmers of North Dakota. The NDAES has travelled many roads toward that elusive goal—breeding improved varieties of crops, developing better livestock, encouraging farmers to diversify production and to adopt more modern agricultural and business methods, and so forth. At times the roads followed proved to be dead ends. Sometimes the NDAES set out on the wrong road, either because it misread the signs or because it could not foresee pitfalls up ahead. Successes and failures are both part of the Station story.

The pursuit of income stability is an important theme of Station activities, but it is not the only one. The NDAES has also periodically regulated a variety of products used by farmers and other consumers, attempted to enhance rural living and strengthen rural families and communities, and encouraged economic development that transcends agriculture as such. These activities, too, are part of the Station story.

The NDAES has seldom pursued its many activities in an atmosphere of total agreement or consensus. Station researchers are seldom completely united on matters of science or public scientific policy. The people of North Dakota, who provide most of the funding for the NDAES, are often even more fragmented. The first half of North Dakota's life as a state was characterized by bitter political struggles between agrarian radicals and conservatives. The NDAES could not escape those struggles. Indeed, it was caught up in them, with especially detrimental consequences in the 1890s, the 1910s, and the 1930s. The NDAES is a political as well as a scientific institution, and so it should be as a tax-supported entity in a democracy. That is another part of the story that follows.

The Station has been subject to many changes over the first century of its existence, but it has also been characterized by a remarkable stability. The value system of the scientists has changed little, if at all. They are believers in human possibility and progress; idealists in some ways, though they think of themselves as practical people. They are people of integrity; they have always told North Dakota the truth, even when that truth was neither pleasant nor welcomed. Most of all, they have been devoted to serving the state and its people. "Our purpose is to serve"—the exact words are those of entomologist Ted Schultz. The idea that underlies them has animated most of the men and women of the NDAES throughout its history. Mainly, the story that follows is the story of that service.

The project that culminated in this book had its genesis in 1982, when Director H. Roald Lund of the NDAES approached Dean Archer Jones of the College of Humanities and Social Sciences about a potential author for the centennial history of the Station. Jones suggested me for the assignment and Lund and I quickly agreed that I should undertake it.

At that time Lund stated his desire to see what he called a "true history" of the NDAES. The problem is that in history truth is in the eyes of the beholder, and some beholders may see more truth here than will others. "*Our Purpose is to Serve*" largely reflects my judgment of the most important developments and the most crucial events in the history of the NDAES. Some readers will wonder why I covered some things and not others, or why I mentioned some people and seemingly ignored their colleagues. These are matters of editorial judgment, and I will stand behind my decisions, but I will also be the first to admit that my judgment is not necessarily inerrant.

I needed to exercise editorial judgment in order to capture the essense of the Station experience. Had I attempted to do justice to all of the people and activities involved in a diverse and complex institution such as the NDAES I would have produced an involved and compendious tome that would be of little interest to most readers. Instead, I tried to draw out and elaborate on a handful of key themes. Doing so necessitated that I exercise a good deal of selectivity and that I simplify the complicated story of a complex institution. Throughout I have attempted to write a "true history" in a sense of a history that is faithful to the record, that seeks fairly to understand what people did and why, and that is honest with the author and his readers. I hope that my readers will agree this is a true history in accordance with those standards.

One of an author's most pleasant tasks is to acknowledge some of those who helped him produce a book. I profited from the comments of several readers who contributed their valuable time to me by reading portions of the manuscript. The late Larry Remele, taken from us prematurely in June of 1988, read the first six chapters. Those chapters benefited from his wisdom and acuity. The last three chapters, unfortunately, did not enjoy his ministrations. If they are weaker than the others, that is probably why. My colleague and friend Larry Peterson read the entire manuscript with the same care, compassion, and conscience he exhibits in all of his activities. Richard Overfield, of the University of Nebraska, Omaha, and Donald Green, of Central State University, also read portions of the manuscript; the former was especially helpful in identifying some serious weaknesses in chapter five. Roald Lund and Bill Dinusson read the entire manuscript and saved me from several potentially embarrassing errors. Don Anderson and Ross Wilkinson read and commented on the last three chapters.

John Bye, archivist at the Institute for Regional Studies, helped me find relevant material in that collection and gain access to some important sources at the State Historical Society of North Dakota and at the Libby Library of the University of North Dakota. Bye and Gary Moran helped me identify and reproduce the photographs that are included here. As always, Michael Lyons was a supportive friend and colleague. I am also grateful to the many people who agreed to be interviewed by me for this centennial history. All helped me gain some perspective on the NDAES and its activities; I am sorry only that I was unable to interview more people and to cite all of the people who shared their time and thoughts with me in the book.

Janis Kirsch, history secretary, did her customary excellent professional job of typing and retyping the manuscript, bearing last-minute revisions and even manic moments with her usual good humor.

I wish to thank Archer Jones, who suggested me for this assignment, and Roald Lund, who not only chose me to do the history of the Station but who provided me with encouragement and support every step of the way. Most station histories are done in-house by amateur historians. I believe that Lund's decision to have a professional historian do the history of the NDAES was wise, forward-looking, and, in some ways, courageous. I hope the readers of "*Our Purpose is to Serve*" will agree.

My children, Elizabeth and Mark, helped give me a sense of the future and the importance of preparing for it. As always, Karen was my anchor and foundation. I can't do my best without her.

ABBREVIATIONS USED IN THE TEXT

ASE	American Society of Equity
BFA	Better Farming Association
GNDA	Greater North Dakota Association
NCIA	Northwest Crop Improvement Association
NCI	Northern Crops Institute
NDAC	North Dakota Agricultural College
NDAES	North Dakota Agricultural Experiment Station
NDBA	North Dakota Bankers Association
NDSU	North Dakota State University
NFMA	Northwest Farm Managers Association
NPL	Nonpartisan League
OES	Office of Experiment Stations
SBHE	State Board of Higher Education
USDA	United States Department of Agriculture

1
BEFORE THE BEGINNING

What wealth and prosperity is in store for . . . [the] farmer when he shall, by means of scientific agriculture, double the yield of his fields. . . and . . . when he shall diversify his crops and combine stock-raising, dairying, and the other valuable adjuncts of farming with his yearly wheat ventures!

Dakota Territory Promotional Pamphlet, 1887

Years will come of abundance and years will come of disaster, and between the two the people will be prosperous and unprosperous, and the thing to do is look the question squarely in the face.

John Wesley Powell to the North Dakota Legislature, 1889

In its organizational meeting on May 15, 1890, the Board of Trustees of the North Dakota Agricultural College and Experiment Station created, on paper, an agricultural experiment station. For organizational purposes, the board appointed Fargo physician Samuel T. Satterthwaite as temporary director, and local agricultural pioneer James Holes as superintendent of the experimental farm. Though the college and experiment station were new institutions in North Dakota, both by 1890 were established, federally sponsored agencies. Indeed, both were seen as parts of the birthright of North Dakota and the other commonwealths that joined the Union in 1889.[1]

I

Agricultural colleges and experiment stations were tangible products of the increasing interest of the federal government in scientific agriculture in the late nineteenth century. During the Civil War, Congress took advantage of the temporary absence of obstructive Southerners to pass the Morrill Land-Grant

College Act. That piece of legislation gave each state thirty thousand acres of federal land for each of its United States representatives and senators for the purpose of supporting colleges that would discover and teach the principles of scientific agriculture. Though a few prototypes of agricultural colleges existed in 1862 when the Morrill Act was passed, it provided the key stimulus for the creation of a system that includes nearly seventy institutions today.[2]

Federal support for experiment stations—institutions designed exclusively to conduct agricultural experiments and reveal scientific principles related to farming—came more slowly. The first agricultural experiment stations were in Germany. Justus Liebig made that country the world leader in the field of agricultural chemistry and the German stations concentrated their efforts in that area of investigation. In 1875 the state of Connecticut created the first government-supported experiment station in the United States. Within ten years there were about a dozen state-sponsored agricultural experiment stations in this country, most of them little more than chemistry shops occupied mainly with analyzing commercial fertilizers and animal feeds.

The early experiment stations were relatively humble and limited establishments, but their supporters in and out of government saw sufficient promise in them to warrant federal patronage. In 1887 Congress passed the Hatch Act, providing an annual appropriation of fifteen thousand dollars for each state that created an agricultural experiment station. Congress decreed that these stations must be attached to land-grant colleges.

The assumptions and motives of the people who created this edifice of education and experimentation were complex. Policymakers certainly recognized that, despite the rapid industrial development the United States was then undergoing, agriculture remained the cornerstone of the American economy on which all else to a great extent rested. Thus the maintenance and advancement of agriculture was a wise national policy. That agriculture should be maintained and advanced in this particular manner reflected the developing attitude that science and education were surer means to individual and social advancement than were the faith in God and moral behavior on which Americans—and others in the Western world as well—had traditionally depended. The idea that education and scientific research were proper domains for government activity was held especially by the Whig party and by its political heir, the Republican party. These political organizations consistently favored federal support for educational and scientific initiatives that they believed would help individuals improve themselves.

This "Whig-Republican" conception of the proper role of government had both liberal and conservative facets. On the one hand it contemplated a degree of activism far bolder than the Jeffersonian conception that dominated American political thinking for most of the first half of the nineteenth century. On the other hand it carried the assumption that any number of social, technical, or economic problems could be solved by individuals themselves if they had proper guidance and training. This essentially conservative notion seemed unrealistic to large numbers of farmers in the late nineteenth century. Already vulnerable to the caprice of nature, increasingly commercial farmers also found themselves dependent on market forces they could not control, and on bankers, railroads, implement manufacturers, and middlemen whose charges they could not influence. The Grange, the Farmers' Alliance, and later the Nonpartisan League were all founded on an assumption very different from that underlying the government's agricultural education and research establishment. All of these groups believed that farmers, as individuals, could not solve the basic problems of their economic lives through education or technical adjustments. To agrarian protest groups, political action to change the economic arrangements affecting agriculture was the best means of solving farmers' problems. This basic difference in underlying assumptions proved to be the cause of a good deal of friction between the colleges and experiment stations, on the one hand, and politically and economically oriented farm organizations, on the other.

A related assumption that underlay the creation of the publicly sponsored scientific establishment was that farmers should become more professional. As the nineteenth century proceeded and as the United States was transformed from a relatively simple, agrarian society into an increasingly complex, urban and industrial one, standards of value were altered in a subtle way. Virtue came to be decreasingly associated with who one was and how one behaved and increasingly connected to what one did. More and more, society honored professionals—those with a particular expertise acquired through specialized training—and derogated those who lacked professional standing, regardless of their personal attributes. No group's status was more depreciated in this process than was that of farmers. In the early Republic farmers were celebrated as social paragons simply because of who they were and what they did. As American society became more complex, sophisticated, and urban, however, farmers found themselves being portrayed increasingly as hayseeds and yokels. The deprecation of farmers intensified at least in part because their occupations seemed to demand no special intelligence, skill, or training.[3]

The publicly sponsored research and education system reflected these emerging attitudes in two ways. First, the agricultural colleges and experiment stations were staffed by those with rising social status—specially trained experts commanding bodies of esoteric technical knowledge. And second, the purpose of these institutions was largely to professionalize farmers and thus to elevate their social status. As W. O. Atwater, the first director of the USDA's Office of Experiment Stations, suggested, the scientific expertise offered farmers by the stations "elevates farming as a profession, and . . . shows . . . that it is a profession in which brains can be used with profit." That supporters of the system hoped to elevate farmers and help them retrieve social respect was laudable. That they made concessions to narrow urban standards of value in the process was unfortunate, if perhaps inevitable.[4]

Other, more concrete considerations also contributed to the creation of these educational and research institutions. By the time of the Civil War far-sighted policymakers could perceive a crisis in soil fertility. Yields were declining in areas of the country settled for as little as twenty or thirty years. Suddenly the American agricultural domain, apparently so vast and inexhaustible a bare generation before, seemed in danger. Perhaps, policymakers thought, science could help Americans escape the consequences of their folly.

As Americans frittered away so much of their natural bounty they looked, as they always had, to the West. There they saw not more of the well-watered, abundantly timbered lands they were accustomed to, but vast, treeless prairies, rugged mountain ranges, and trackless desert wastes. Here were special challenges even for people used to overcoming challenges, and perhaps science could help find answers to how American agriculture could be extended into a strange new land.

To solve agriculture's problems by helping people help themselves, to educate and professionalize farmers, to maintain America's agricultural greatness, to rebuild soil fertility, and to meet the challenge of the West constituted a heavy load for the infant educational and scientific system to carry. Perhaps it was too heavy, because some things could not be done quickly and others could not be done at all. But this was the package of assumptions with which the system was burdened and which it bore in North Dakota as elsewhere.

II

North Dakota was part of the strange new land of the West for which the experiment stations seemed especially appropriate. Broadly speaking, the state is part of the Great Plains region of North America which stretches from the prairie provinces of Canada into the northern states of Mexico. But within the state there are great variations in topography, soil type, and climate.[5]

Approximately 10 percent of the state is level prairie, formed at the bottoms of prehistoric lakes. The largest of these was Lake Agassiz, which formed the Red River Valley that straddles the Minnesota border on the eastern edge of the state. When settlers came the level prairie was tall-grass country, a reflection of relatively abundant moisture and uncommonly rich soil.

Drift Prairie comprises about 40 percent of the state, including much of the eastern, central, and northwestern portions. This is a glaciated region characterized by gently rolling hills and dark chernozem soils. The natural vegetation here runs from tall to short grass, depending mainly on the availability of moisture.

Most of the western portion of the state is called the Missouri Slope because it is drained by the Missouri River. The glaciated portion of this region resembles the Drift Prairie in soil type and topography. The southwestern portion of the Slope, however, was not touched by glaciers, and it is characterized by chesnut soils and steep-sided hills, bluffs, and plateaus. The natural vegetation in this region, which covers about 40 percent of the state, is short grass.

North Dakota provides one of the best examples in North America of the continental climate. Fifteen hundred miles from any climatically significant ocean or sea, North Dakota enjoys less of the moderating effect of bodies of water than any other state. Consequently, winters are extremely long and are colder than those in any of the other contiguous states. Summers are relatively short but can be quite hot. The wind is a major climatological factor in all seasons. Cold winters, short growing seasons, and wind have each presented difficult challenges to farmers and to agricultural researchers in North Dakota.

The state's distance from major bodies of water also affects the moisture supply. Some moisture comes from the west, but most Pacific fronts are relatively dry by the time they reach North Dakota. Most of the state's moisture comes from the Gulf of Mexico, but the distance it must travel dictates that the states to the south are progressively better watered. Only a thin strip of eastern North Dakota receives as much as twenty inches of moisture per year, rendering most of the state semi-arid. To some extent this apparent aridity is deceptive. As early as 1873 government researcher John Wesley Powell pointed out that, because of the nature of the soils and the rate of evaporation, an inch of rain was more fully utilized by crops in what was then Dakota Territory than in the Southern Plains. He also pointed out that over 70 percent of moisture in Dakota fell during the growing season, when it was most needed for agricultural purposes. Still, water has been a constant concern for farmers and for researchers throughout the state's history, whether there was too much of it, as has often been true in the Red River Valley, or too little of it, as has periodically been the case everywhere else in North Dakota.[6]

As part of the Great Plains, North Dakota shared the stigma of the "Great American Desert" appellation fastened on the region by explorers and cartographers early in the nineteenth century. During the 1870s, however,the possibilities of the region began to be revealed. The impressive production and spectacular scale of the "bonanza" farms—huge wheat-raising operations of thousands of acres encouraged and indirectly supported by the Northern Pacific Railroad—drew the interest of settlers to the Red River Valley. The bonanza operation directed by Oliver Dalrymple about twenty miles west of Fargo at Casselton became the focus of intense national media attention and the object of visits by such distinguished guests as President Rutherford B. Hayes and a British royal agricultural commission. Though visitors to the area continued to find it strange and aesthetically unappealing, they began to emphasize its possibilities rather than its limitations, speaking of it more often as a garden than a desert.[7]

By the early eighties the "first Dakota boom" was well under way.Settlers poured into the Red River Valley and farms began to dot the Drift Prairie region as well. Open-range cattle ranching, pushing northward inexorably from Texas, thrived in the southwestern part of the state, attracting such romantic figures as the French Marquis de Mores and the young Theodore Roosevelt. High prices for hard red spring wheat and the apparently insatiable demand by Easterners for beef seemed to indicate a bright future for the would-be state. Boosters of the territory and boomers of its towns, who had staked their economic lives on the expectation that northern Dakota would be a new Iowa or Wisconsin, spoke confidently of the future of this latest and greatest frontier.[8]

As quickly as it had materialized the boom evaporated. Severe winters, dry summers, and declining prices drove many open-range ranchers into bankruptcy and forced others to alter their operations dramatically. Drought, wheat stem rust, and low prices deviled grain growers, forcing many out of business and others into debt. By 1890 nearly half of the farms in North Dakota were mortgaged. The rural population receded and the towns shrank.

Optimism remained the most dependable commodity in the area, and as late as 1889 a speaker could predict with a straight face that Bismarck would some day be the center of Western civilization. But the end of the boom stimulated some thoughtful reexamination among many of the less fantastic boosters of North Dakota. One clear problem was that of water. Champions of the region preferred to believe the pseudoscientific theory of Dr. Samuel Aughey of the University of Nebraska that rain would follow the plow. Other observers were more realistic. John Wesley Powell warned the first state legislature that wet periods would alternate with dry periods in most of the state. In the former the people would be "prosperous" and in the latter they would be "unprosperous." The late eighties seemed to indicate that Powell was right.[9]

Then there were the problems of wheat monoculture in the state and the wasteful methods employed by many farmers. The concentration on one crop placed the health of North Dakota's economy at the mercy of a fluctuating world market and raised the possibility that some natural phenomenon—such as a virulent outbreak of stem rust—would drag the state down to ruin. Even at best wheat would not make the state prosper if poor methods returned low yields and diminished the fertility of the soil. In 1887 even a promotional pamphlet for Dakota Territory suggested that future greatness was not automatically assured but might depend on better farming methods and agricultural diversification.[10]

Other problems were equally vexing and perhaps even more intractable.North Dakota was remote from major markets and population centers and distance diverted potential profits into freight and handling charges and raised the prices of imported products. The railroads farmers depended upon lacked competition and consequently charged high rates. Grain elevators, most of which were closely associated with the railroads, downgraded the grain they bought and upgraded what they sold. Capital was scarce and interest rates were high. Merchants, implement dealers, and other farmers of the farmers charged high prices because of the costs of distance and capital and, frequently, the lack of competition.

For some boosters of the infant state the experiment station was one of the quick-fix miracles that would solve these emerging problems. Perhaps science could help the state overcome the unpleasant facts of a difficult environment and a capitalist economy.

III

It was ironic that people should place their hopes on agricultural experiment stations as those institutions existed in 1889, when North Dakota entered the Union. The stations themselves were immature institutions struggling for self-definition and even survival.

In most states they quickly became embroiled in politics. Disputes arose over where the stations should be located and who should control them. Federal funds drew the attention of state politicians, who usually viewed the local experiment station as just another source of patronage. Stations were often caught in the middle of factional and party battles.

Political interference disrupted the work and intensified the problem the early stations had with defining what their roles should be. Were they to teach, experiment, generate revenue, undertake chemical analysis, maintain model farms, or engage in some combination of these activities? Should experimentation be "applied" or "pure" or both? How was the work to be planned and who would do the planning? And how was the work to be carried out?

This basic problem of role definition was exacerbated by the nature of the support for the early stations. While scientists, policymakers, agricultural editors, Western boosters, and railroads had favored federal sponsorship of agricultural experimentation, few actual farmers had done so. This raised the question of whether, and how, farmers should be involved in the work or should be consulted about it. It also raised the issue of how support for agricultural science—which farmers commonly stigmatized as "book farming"—could be generated in the rural community. Farmers might support experimentation if they saw it benefited them, but that dictated an emphasis on short-term, practical work at the expense of more fundamental and, potentially, more valuable work. And how were farmers to be acquainted with the work? The Hatch Act required each station to publish at least four bulletins per year. That was fine for farmers who could read *and* knew English *and* could understand technical prose *and* would take the time to master a publication *and* could get on the station mailing list. But how many farmers met all of those requirements?

Another problem faced by the experiment stations derived from the fact that Congress had created institutions based on weak scientific foundations. Most of the disciplines represented on the staffs of the early stations were underdeveloped and immature, and few of them even had the theoretical underpinning to justify their separate and independent existence. Too little was known about the basic facts of plant, animal, and insect life to provide anything like an adequate foundation for most of the agricultural sciences. Chemistry was the only discipline represented in the stations that would be defined by modern scientists as well grounded and mature. Botany and zoology both existed, but they were barely beginning to move beyond the natural science tradition of observation and classification. Such areas as microbiology, genetics, parasitology, and entomology were vague, inchoate, and undeveloped. This problem was intensified by the fact that the sharp increase in demand for scientists occasioned by the onset of federal support spread whatever talent there was even thinner.[11]

The weakness of science and the shortage of scientists resulted, in the words of W. O. Atwater, in "inexperience and superficiality" in early station work. Poorly trained people undertook poorly planned and poorly controlled experiments. Scientists in one state duplicated the work of scientists in the next without even knowing they were engaged in duplication. Simple practical experiments proved challenging enough, and the limitations of scientists and science effectively precluded more fundamental work. About the best the stations could do in the early years was to demonstrate the highest standards of contemporary farm practice. At worst, through ignorance or carelessness, they promulgated misleading or wrong information as scientific fact, hurting farmers and diminishing their own credibility in the process.[12]

The North Dakota Agricultural Experiment Station could not avoid the problems endemic to the larger enterprise of which it was a part. Political interference deviled the Station for much of its first half century, disturbing the scientists and threatening to disrupt the work. It was a long and difficult process to interest farmers in the work and communicate effectively with them. Scientists in North Dakota, as elsewhere, struggled with their own limitations and those of their disciplines in their efforts to contribute to agriculture. Eventually the Station became an institution of confident and competent people, respected in the state for its service to North Dakota agriculture. But this position was not achieved immediately and indeed was gained only through a long and painful process.

IV

In the early days of statehood those North Dakotans who thought much about it saw mainly promise in the idea of an agricultural experiment station. There is little doubt that, had they understood all of its limitations, they would have welcomed it anyway. If it achieved nothing else the Station would bring an annual federal payment of fifteen thousand dollars to the state. The same economic marginality that made North Dakotans hopeful about what the station might do dictated that they would embrace it even if, in the end, it did very little.

The questionable economic prospects of North Dakota made the competition among the principal towns for state and federal institutions especially keen. A facility that employed local people and brought trade to local businesses might assure future prosperity or at least survival for a struggling community. For the competing towns the importance of an institution to the state's future was distinctly less significant than the number of dollars it could be expected to pump into the local economy. Hence, towns rolled logs, scratched backs, and made bewildering and shifting alliances in their attempts to capture the choicest plums.

In 1883, when the Dakotas were still united as Dakota Territory, the legislature chose Fargo as the site for an agricultural college. Characteristically, however, it refused to appropriate any money to bring the legislation to life. Fargoans were ambivalent about the prospect of a land-grant institution in their city in 1883, and that ambivalence endured in 1889 when the North Dakota Constitutional Convention got down to the business of passing out institutions. Some feared that an agricultural school would compete with Fargo College, part of an impressive effort by the Presbyterians to civilize the Great Plains, and others believed an insane asylum or penitentiary promised greater economic benefit to the town. Agricultural students seemed rather hypothetical, while everyone knew there would always be criminals and lunatics in abundance.[13]

Fargo's lukewarm attitude about being the land-grant site was manifested in pressure on the town's delegates to the Constitutional Convention. Delegate Samuel Roberts reminisced in 1921 that many in the city would have preferred "an Indian school, or a school for the deaf and dumb, or a school of science, something big!" Unfortunately for Fargo's aspirations the choicest plums went elsewhere, and the town's delegates were able to secure only the agricultural college that had been promised six years before. While the city obviously wanted something, there is no question that Fargo was able to contain its enthusiasm about being made the home of the agricultural college and experiment station. As Roberts concluded, "Fargo actually had the Agricultural College thrust upon it, its wishes to the contrary."[14]

In retrospect, the decision to place the College and Station at Fargo, dictated as it was by an exercise of political power, appears to have been an unwise one. The location of the Station in the Red River Valley meant that field experiments were of questionable validity for most of the state. Had agricultural scientists been consulted they might have urged placement of the Station at a Drift Prairie site, such as Valley City or Jamestown. The location of the Station on the Minnesota border also made it remote and isolated from much of the state. A more central location would have facilitated contacts between the Station and its agricultural constituency.

Another problem related to Fargo's perception of itself. Fargo was the biggest town in the state, and it grew into the major city. The attitude of many of its citizens toward the new agricultural research and educational institution reflected a degree of urban snobbery and condescension. The people at the College and Station did not usually have a sense of belonging in Fargo, in part because of the city's attitude, in part because of their own rural backgrounds, and in part because the institution was built on the outskirts of town. The College and Station were, and would remain to some extent, *in* Fargo without being *of* Fargo.

The state's perception of Fargo compounded the problem. Many people in North Dakota have traditionally viewed Fargo with unconcealed fear and hatred. Some of this attitude derived from the alien urbanity of the town and some of it arose from the fact that Fargo was closely tied to the Minneapolis business interests that many people in the state blamed for the farmers' problems. This business connection and the staunch and predictable conservatism of the city's leading figures in times of agrarian protest and unrest did not endear Fargo to many in the agricultural community. Perhaps inevitably, some of the fear and distrust of Fargo carried over to the land-grant complex located there, making it that much more difficult for the institution to serve the state.

The unfortunate ramifications of the decision to locate the College and Station in Fargo notwithstanding, the deed was done, and the Board of Trustees turned to the job of bringing life to the institution. In the summer of 1890 the Board hired Claire B. Waldron, a horticulturist, and on October 15 it confirmed as permanent president of the College and director of the Station Dr. Horace E. Stockbridge. Confirmed at the same time were two of Stockbridge's appointees, chemist Edwin F. Ladd and botanist Henry Luke Bolley. At that point, Stockbridge and his tiny staff embarked on the difficult and exciting task of creating an agricultural experiment station for North Dakota.

2

EARLY DAYS

We wish to impress upon the attention of the farmers of the State that this institution belongs exclusively to them, that it receives its income for the sole purpose of being expended in their aid, that in its work their desires and interests will be constantly furthered, and that their interest, assistance and suggestions are earnestly desired, solicited and expected.

H. E. Stockbridge, 1891

Just now my principal business is that of doing my share of what is necessary to keep the Agricultural College from being absorbed by the Populists.

James B. Power to Andrew Burke, 1893

The task that confronted Horace E. Stockbridge and his young colleagues at the new North Dakota Agricultural Experiment Station was imposing, but also exciting. They were given an opportunity that few people ever have—to create an institution, establish its traditions, and set it on paths that it might follow long after the trailblazers were dead. It was a job that demanded strength and vitality, but also care and wisdom. It was a challenge of complexity, subtlety, and no small amount of risk, but one that Stockbridge and his colleagues embraced eagerly.

Stockbridge set his course by the guiding star of service. "We wish to impress upon the attention of the farmers of the State that this institution belongs exclusively to them . . . ," he assured the readers of his first public report as director, "that in its work their desires and interests will be constantly furthered." Directors and staff members have come and gone, research emphases have shifted, scientific practices and standards have changed, and the means of communicating results have been revolutionized, but the commitment to service made in the early days has never wavered.[1]

The degree to which the Station would be able to serve the state depended largely on how it addressed the pressing tasks of organization and popularization.

The first entailed everything from erecting buildings and preparing land to choosing personnel and planning research. The second involved the equally intricate and more subtle problem of gaining the attention of farmers, capturing their support, and stimulating their enthusiasm. Stockbridge and his colleagues could not foresee that fulfilling these tasks would be complicated by a devastating agricultural depression and by intense factional and partisan political interference in the operation of the institution. In light of their difficulties it was a tribute to their energy and devotion that the early scientists were able to organize the station and take some tentative steps toward popularizing its work.

I

The burden of providing early organization for the Station fell largely on the new director, Horace E. Stockbridge. Only thirty-three when he received his appointment, Stockbridge already had a depth of training and a range of experiences that seemed to make him well qualified to address the challenges at hand. Stockbridge came from a distinguished Massachusetts family with roots in the early colonial period. His father was an accomplished educator who served as president of the Massachusetts Agricultural College. After receiving his doctorate in agricultural chemistry from the University of Goettingen in Germany, Stockbridge served in the Japanese Imperial College of Agriculture and Engineering and as director of the Indiana Experiment Station. His frequent changes of position perhaps indicate how numerous were the opportunities open to trained agricultural scientists a century ago. But Stockbridge also had a highly developed sense of professionalism and an aversion to political interference in station management. That tendency, coupled with his sympathy for farm protest groups, led to his dismissal as director of the NDAES in 1893, but he went on to a long and creative career in education, journalism, and farm organization leadership.[2]

The trustees had already hired Claire B. Waldron when Stockbridge came on board. Twenty-six when he gained his appointment, Waldron had received a bachelor's degree and some graduate training at the Michigan Agricultural College in botany and horticulture. As was true with the other early scientists, he was a jack-of-many-trades who added forestry and entomology to his speciality of horticulture. Indeed, Waldron's first assignment at the Station was to rush to Cavalier and Bottineau counties to fight an outbreak of grasshoppers, which he did with the use of tar pans. C. B. Waldron served the institution for half a century, including a stint as dean of Agriculture. His brother, Lawrence Root Waldron, joined the staff in 1899. The younger Waldron, who also served for over fifty years, was the Station's first wheat breeder and a man who enjoyed great respect both in the state and in the larger scientific community beyond.[3]

The position of chemist was the premier post in most early stations, both because chemistry was the most advanced of the early agricultural disciplines and because chemical analysis was often their main area of concentration. For this critical position Stockbridge secured Edwin Fremont Ladd, a thirty-year-old assistant chemist from the New York Agricultural Experiment Station at Geneva and a graduate of the University of Maine. At Geneva, one of the leading stations in the country, Ladd had served with distinction under such leaders in the field as Stephen Babcock and E. Lewis Sturtevant. Ladd was apparently attracted by the challenge of helping shape a new institution in a new country. He became the most famous and popular member of the staff in the early years, largely because of his pure food advocacy and his controversial milling and baking studies. Self-confident and ambitious, Ladd parlayed his station work into the presidency of the Agricultural College and election to the United States Senate.[4]

The other two researchers hired by Stockbridge in 1890 had been his junior colleagues at the Indiana Station. Theries D. Hinebauch, the thirty-year-old veterinarian, was a Michigan Agricultural College graduate. Stockbridge hired Henry Luke Bolley away from the Indiana Station as botanist and plant pathologist. Merely twenty-five when he came to Fargo, Purdue-graduate Bolley had already done extensive work on the problems of wheat rust and potato scab. In terms of economic benefits to the state of North Dakota Bolley was the most important of the early scientists. Any one of his three major contributions—advocacy of the formaldehyde treatment for smutted seed, discovery of the cause of flax wilt, and recognition of the barberry bush as a host for wheat stem rust—surely saved the state enough money to more than pay for the entire operation of the Station during the sixty years Bolley worked there. Bolley was renowned as a plant explorer, and his early trips to Russia resulted in the introduction of rust resistant strains of hard red spring and durum into the NDAES wheat-breeding program. It was Bolley's fate to become a living anachronism, owing mainly to his loyalty, at a time when most of the Western scientific world embraced Mendelian genetics, to the Lamarckian notion that acquired characteristics in plants and animals could be passed to offspring. His scientific heresy, along with his prickly egotism and outspoken political radicalism, alienated him from many younger colleagues at the Station. Nevertheless, he was regarded with great affection in the state, largely because his devotion to the farmers was so pure and selfless.[5]

The early members of the Station staff shared several characteristics. First, all but Ladd came from either Michigan or Purdue. Competing on a job market in which the demand for scientists far exceeded the supply, Stockbridge depended on his own contacts and the personal networks of friends to build the Station. The fact that Stockbridge was the only scientist with a terminal degree also related to the job market. Ph.D.'s in the agricultural disciplines were scarce and expensive. Stockbridge had no alternative to hiring promising people with lower levels of formal education. The most striking characteristic of the staff was its youth. Only thirty-three himself, Stockbridge was the senior member of a group of young people. In large part the youth of these scientists is explained by the newness of the agricultural disciplines themselves and by the greater likelihood that people with shallow roots and minimal family ties will move to a new institution in a new place. The youth of the staff led to some missteps and errors in judgment in the early years, but it also imparted a high level of excitement, vitality, and congeniality to the early station experience. The youth of the Station scientists in 1890 resulted in many decades of devoted service to the state from Bolley, Ladd, and C. B. Waldron, who grew up with the station and, ultimately, grew old in it.

One of the attractions that brought this young group to North Dakota was the prospect of building an institution where none existed before. As they contemplated this task from their temporary quarters at Fargo College, Stockbridge and his colleagues began to realize the magnitude of the job of building an experiment station literally from the ground up. Procedures had to be agreed on, the relationship between the College and the Station had to be defined, land for experimental purposes had to be acquired and prepared, quarters had to be erected, equipment had to be secured, and, most importantly, a general research mission had to be delineated and specific projects undertaken. In their efforts the early scientists were guided to some extent by the experience of other states, and sometimes the USDA's Office of Experiment Stations was solicited for advice. While the OES tried to be supportive it left as much as it could to state initiative, especially in the case of issues with political implications. Stockbridge and his colleagues were generally on their own, guided by their sense of correct procedure and outstanding needs and their commitment to service, but always constrained by the economic and political climate of the state.[6]

Of the problems that confronted Stockbridge and his colleagues the question of the relationship between the College and the Station proved the easiest to solve. In most states land-grant colleges preceded the experiment stations, which were grafted onto them. This frequently contributed to friction, resentments, and unclear definitions of authority. In North Dakota, the Station and the College were created together, grew together, and, initially, had an identical staff and administration. Stockbridge was both president of the College and director of the Station, and these positions were not separated for nearly a quarter century. Given the miniscule size and overwhelmingly agricultural orientation of the student body in the first years, this arrangement was reasonable, and the early scientists found it desirable. Bolley, for example, argued in 1894 that "the work of the college and experiment station, though having very different ends in view . . . are fittingly associated. Much of the work accomplished being of co-ordinate advantage to both, instruction, appliances, books and apparatus are thus used in common for the two lines of work."[7]

One problem with the close relationship between College and Station was that it complicated finances, and Hatch Act funds inevitably drifted into the college account, to the substantial consternation of the USDA. This fund mixing was illegal, but in these early years it was necessary. It is highly unlikely that the

College would have been able to function during the depression of the nineties had it depended solely on state appropriations. The Office of Experiment Stations believed that the cause of science would be better served if stations and colleges were strictly separated, and OES officials urged that course on North Dakota and other reluctant states. When a clearer separation was finally implemented in 1914 it was the cause of much animosity and conflict and must be viewed as incomplete and only partially successful. Over the years the College and Station have become more distinct entities, but even today they are more closely intertwined than is the case in most states.

Another pressing problem was that of securing facilities to undertake experimental work. Early in 1891 the legislature appropriated funds for a college building that promised eventual escape from the cramped quarters being leased at Fargo College. The legislature also appropriated money for the purchase of a section of land one mile north of Fargo. The College and Station did not come into possession of the land until April 18, 1891. At that time Stockbridge found that much had to be done. Roads had to be graded, fields had to be laid out, and quarters for men, machinery, and animals had to be erected. Much time was lost because men and teams had to come out from town every morning and return every night. The use of workers for the construction of buildings further delayed field preparations. Two quarter sections of the land had been cropped by renters for a dozen years and suffered from diminished fertility and the prominent presence of "noxious weeds." In addition, the section included a shallow lake of about fifty acres that the staff hoped to drain when time and money became available. The scientists were frustrated by the delays in getting field work under way, and others in the state might also have wondered when some tangible returns would be forthcoming from the Station. The first year passed into the second and Stockbridge wearily reported to the governor that "the present year, as was its predecessor, has . . . necessarily been devoted largely to organization . . . and the effecting of such permanent arrangements as must necessarily precede the actual development or inauguration of systematic investigation."[8]

While housekeeping matters were being settled, issues that would be of more enduring significance were being addressed. These involved the general and the specific directions in which research at the Station should go.

In the early years of the experiment stations there was a vigorous debate between the proponents of "pure" research and the supporters of "applied" research. These were, to some extent, artificial distinctions, because applied or practical research sometimes resulted in the discovery of basic scientific principles while pure or theoretical research promised ultimate practical applications. Still, the distinction was a meaningful one to early agricultural scientists, both at the NDAES and in the country as a whole.

The essential argument of the supporters of pure research was that only through this emphasis would basic principles of enduring importance to agriculture be discovered. W. O. Atwater, the first director of the OES and a proponent of the theoretical emphasis, put the case succinctly when he argued that "the ultimate success of the stations will depend upon the discovery of principles." This position was represented at the NDAES by H. L. Bolley. "While abstract investigation for abstraction's sake is hardly to be recommended in American institutions," Bolley told his colleagues at a convention of the Association of American Agricultural Colleges and Experiment Stations, "the economic purpose never alone rules the work of a scientific investigator." Bolley argued elsewhere that the primary question stations must ask themselves was "how much you have augmented the aggregate of working principles of agriculture."[9]

Atwater, Bolley, and other proponents of pure research were also interested in elevating the new disciplines of scientific agriculture. Early station researchers were held in low regard in the American scientific community. Not only were they servants of the farmer, who was increasingly viewed by an urbanizing society "as a hayseed, [and] a figure of fun and bland contempt," but they were also operating in new disciplines that lacked tradition, clear direction, and established principles. When they emphasized practical research, station scientists only increased the contempt the more theoretical practitioners in established disciplines had for them. Denigration of station researchers by the scientific elite continues even today, but a century ago it was much more virulent. It is not surprising that many agricultural scientists searched for ways to elevate themselves and their work.[10]

Those who believed that practical research should be emphasized did not deny the points their theoretically oriented colleagues made, but they did appeal for a more realistic understanding of conditions in society and in the agricultural disciplines themselves. The fact was that the stations were public institutions, and their prosperity and even survival depended upon public support. That support would most likely be forthcoming if the stations produced immediate public benefits. They did not have the luxury of undertaking exclusively theoretical investigations that might not result in practical applications for twenty or thirty years. As one supporter of theoretical investigation ruefully admitted, "the great [public] demand is for something 'practical,' by which is meant something whose value is at once apparent, and can be measured in dollars and cents."[11]

The demand for the practical was insistent everywhere, but nowhere more than in new states such as North Dakota. Substantial agriculture had been carried on in North Dakota for barely two decades, and farmers were still casting about for the most appropriate crops and methods on this last frontier. As A. W. Harris, Atwater's successor at the OES put it, "in our newer . . . communities the stations . . . find sufficient employment in pioneer experiments, by which the farmers might be taught what crops they should introduce and what abandon, how agriculture could be more diversified, and how the resources of soils and water could be best utilized and husbanded." The OES chief concluded that "if our stations are for the most part working in these practical lines it is because there is such imperative need on the part of the farmers for advice and direction in those ways which do not involve the discovery of new principles, but only the application of old ones to peculiar circumstances and conditions."[12]

The limitations within the new disciplines and among the new scientists also favored the applied emphasis. Only agricultural chemistry was a mature discipline with a firm theoretical foundation. The other disciplines groped around in ignorance and insecurity, attempting to define themselves and contribute to knowledge. While basic research eventually provided these disciplines with the foundations they needed, their immaturity and incoherence severely limited the number of scientists who could fruitfully and intelligently undertake such work.

Out of these immature disciplines came weak scientists who often lacked confidence in their training. Very few had terminal degrees, and some did not even understand the basic principles of the scientific method. Carrying forward a replicable practical experiment was enough of a challenge for most of these pioneers. Discovering fundamental truths of plant and animal life far exceeded their capabilities. The problem of quality and training was, ironically, made worse by the Hatch Act, which dramatically increased the demand for agricultural scientists. As Wisconsin director W. A. Henry admitted in his presidential address to the Association of American Agricultural Colleges and Experiment Stations convention in 1893, "in calling to this work in a single year hundreds of men it was impossible that all should be properly trained for the work, and many of those selected were not fitted for it either by training or by natural adaptation."[13]

Both social and scientific factors pointed the stations in the direction of practical work of a particular kind: cultivation and feeding experiments, variety testing, classification of local plants, and chemical analysis of feeds, fertilizers, soils, water, and so forth. It was in these areas that the early researchers, with their analytical and observational skills, could make immediate contributions to agriculture and even begin to fill an empirical reservoir on which more theoretical scientists could draw.

Stockbridge was a strong proponent of practical work at the early NDAES. In his first annual report he noted, "we recognize that the institution must be strictly agricultural, and that its work, to meet with public approbation, must be of a positively practical character, . . . and the staff will exert itself as a unit toward the accomplishment of definite practical results of value to the intelligent farmers of North Dakota." But in favoring a practical emphasis, Stockbridge did not propose to create a glorified chemistry shop, such as many of the Southern stations were, or to erect a model farm that would illustrate the best agricultural practices to visiting farmers. He and his colleagues hoped to create a working station that would investigate and experiment on the practical problems faced by North Dakota farmers. Nor did he wish by his practical emphasis to discourage Bolley, the only one of the early researchers who made a substantial contribution to scientific knowledge. There was room for theoretical work, especially when it dealt with problems of importance to the state's agriculture. The formula devised by Stockbridge and his colleagues, which stressed rigorous practical work without excluding theoretical research, has been followed by the Station with little deviation for a century.[14]

The specific areas of research concentration in the early Station can be largely encompassed by one word—wheat. Wheat was the dominant crop in the state and the early research efforts at the Station were mainly directed either at improving the quantity and quality of the wheat produced or at developing crops that could supplement or provide an alternative to wheat. Bolley was brought to North Dakota in part because he had done work on wheat rust. He remembered that Stockbridge told him "that the major line of work" of the botanist would be "the control of wheat diseases." The agriculturist, who managed the Station farm and concentrated on such areas as variety testing and cultivation, fertilization, and weed control experiments, shared this emphasis. "Experiments looking to the improvement of the 'No. 1, Hard' quality of our state's greatest product, wheat, and to not only keeping up the yield, but increasing it, appeals to the interest of every loyal citizen of North Dakota," wrote agriculturist Willet Hayes in 1893.Outside of these departments there was less concentration on wheat, but in no department was it ignored. In Chemistry, for example, there was little initial interest in the crop, but within a few years milling and baking studies were occupying increasing amounts of Ladd's energy and were advancing his reputation as the people's champion.[15]

Wheat was such a dominant crop in early North Dakota that no one at the Station ever questioned this research emphasis. Indeed, even today the Station concentrates more energy on wheat than any other single crop, though wheat does not dominate other crops as it did when the NDAES was founded. In 1895 the USDA estimated the value of the North Dakota wheat crop for the year to be over 23 million. The total value of corn, oats, barley, potatoes, and hay produced on the state's farms was estimated at less than $7 million, and the worth of cattle, sheep, and swine in North Dakota was reckoned at only $9 million. Wheat was king, and its well-being preoccupied most of the 80 percent of North Dakotans who farmed. They were accustomed to growing it, and the spring wheat they produced was the basis for the finest bread flour made in America. By serving wheat the early Station served the public interest in the state to a degree unmatched by most other stations. At the same time early Station scientists as well as state leaders in public affairs recognized that wheat was a mixed blessing for North Dakota. It was truly the staff of life for the state, but concentration on it meant an undiversified agricultural and general economy, seriously susceptible to price fluctuations and natural calamities. As long as North Dakota lived by exporting commodities it would be colonial; as long as it lived by exporting one commodity it would be colonial and vulnerable.[16]

The NDAES could do little about the economic dependence of the state on commodities, but it could and did endeavor to diversify the commodity mix. From the beginning the Station investigated new plant and animal products that, if they were successfully introduced, might diminish the state's dependency on wheat. Along with the ever-present bulletins on wheat there appeared in the early years publications on potatoes, sugarbeets, vegetables, sheep, and dairying. Two of these—potatoes and sugarbeets—eventually became important crops. Early vegetable experiments mainly confirmed the difficulty of growing these crops in the North Dakota environment. As C. B. Waldron wryly noted in one bulletin, "I would not advise farmers to rush headlong into vegetable culture."[17]

Dairy farming became an obsession at the Station and among business boosters of the state, but one in which most North Dakota farmers resolutely refused to share. Supporters argued that dairy farming would stabilize producers' incomes, provide for efficient use of forage, enhance soil fertility, and encourage industrial development. Not until after World War II did the Station reluctantly accept the reality sensed by most farmers—that North Dakota would never become a dairy state—and adjust its research program accordingly.[18]

Although the development of new and existing cash crops dominated the early work of the Station, there was also a concern with the quality of rural life. C. B. Waldron's work with vegetables, fruits, trees, shrubs, and flowers was directed mainly at enriching rural people in a spiritual rather than a material sense. A believer in the positive effects of a pleasant natural environment on the soul, Waldron was truly an "apostle of outdoor beauty," as historian Bill Reid characterizes him. A practical horticulturalist, Waldron never forgot that among the "chief joys of life" were "homes nestled in trees and made cheerful and bright by flowers, as well as tables well supplied with home-grown fruit." Waldron was the early representative of a minor but noteworthy theme that has carried through the history of the Station—that part of the work of researchers properly includes those areas of agriculture and rural life that involve living rather than simply making a living.[19]

II

In addition to the multifaceted job of organizing and defining the work,the Station also faced the task of popularizing it. As a general rule the early stations were welcomed by local boosters but they were not popular institutions with their agricultural constituents. Farmers traditionally saw agricultural science, or "book farming" as they often stigmatized it, as impractical. They masked their insecurity in the presence of scientists with scorn and contempt. They refused to send their children to agricultural colleges, perceiving those institutions as threats rather than handmaidens to farmers. It was not just that they failed to seek the aid of the educated scientist or farmer, they usually fled from it. Of the few farmers who publicly supported the Morrill Land-Grant College Act, or the later Hatch Act, most were "gentleman" farmers who dabbled at agriculture rather than people who depended on it for their livings. As a consequence of this attitude fortunate stations faced indifference in their early years; unfortunate ones confronted hostility.

The North Dakota Station was in the fortunate category because it had mainly to contend with indifference. Public expressions of antagonism to the Station were as rare as expressions of enthusiasm for it. Coming into a new and largely unsettled state with a challenging environment and a relatively young and literate agricultural population, the NDAES was not forced to endure the same negative public attitudes common in older areas. Still, there were problems. North Dakota farmers had not requested the Station that they had received. They, too, felt uncomfortable with agricultural science and scientists. And they wondered what, if anything, the Station could do for them. Then there were the special problems: the size of the state and the scattered settlement pattern, the heavily German and Norwegian ethnicity of the population which made communication difficult and was sometimes accompanied by distrust of government, and the location of the Station on the far edge of the state in a peculiar region and a feared city. Gaining interest in and support for the work was certainly not impossible, but it was not easy either.

Simply by setting the research agenda Stockbridge and his colleagues took a long step toward popularizing the Station's work. It was virtually always the case that applied work found more favor with farmers than did theoretical research. But while farmers usually saw basic work as pointless, they often expected immediately applicable results from practical experimentation. This expectation created pressures on stations to come to conclusions before they were sure of their ground. Bolley reported promising results in his work on wheat and potato diseases, but he cautioned farmers that "economic results in those lines of investigation are necessarily of slow development, coming only after persistent and long continued search for fact which, together with compilations of the work of others, must constitute the basis for experimental operations." Practical work helped create a constituency, but one that was often impatient and demanding.[20]

Then there was the problem, exemplified by Waldron's vegetable work, that research sometimes told farmers what not to do rather than what to do. Another problem was that of applicability. Stockbridge recognized early on that results from the main Station were of questionable validity outside the Red River Valley. Consequently, he concluded that it was "absolutely inevitable, and . . . in every way desirable, that practical field experiments be undertaken in several different localities throughout the state." This dictated either substations or experimentation by cooperating farmers. Since money was unavailable for the former the Station leaned on the latter in its early years. Unfortunately, the difficulty of maintaining scientific standards on working farms meant that the field work done there had a greater value for public relations than for experimentation.[21]

Some problems of farmers could be addressed immediately and the Station offered its assistance regarding these. Stockbridge constantly emphasized that the Station belonged to the farmers and that it existed to serve them. As such, it welcomed farmers' "interest, assistance, and suggestions" and it offered immediate service besides. For example, Stockbridge stressed that "*we gladly receive samples of waters, soils, minerals, crops, or by-products for analysis, determine . . . the character of weeds, identify seeds or insects, the germinating property of seed grains—in fact, seek opportunity to go to assistance of the citizens of the state in any and all directions wherein the application of science may be expected to facilitate or aid practice.*" The open, friendly, and helpful attitude that Stockbridge conveyed worked. Within a few years the Station was receiving hundreds of letters per month from North Dakotans requesting chemical analysis, agricultural guidance, and advice on an impressive array of other subjects, some only vaguely related to farming. The work of the Station was becoming popular, but there was a tradeoff. Every chemical sample analyzed by Ladd, every grasshopper or hessian fly outbreak fought by Waldron, every seed studied by Bolley, and every bit of advice given by agriculturists Willet Hayes and John Shepperd drew the farmers closer to the Station even while it drew the scientist away from experimental work.[22]

There were two other ways in which the early Station popularized its endeavors and built support for itself. The first of these involved farmers' institute work. Farmers' institutes were agencies of adult education that already existed in the older states when the NDAES was founded. Institutes were held in local areas for a day or two, usually in the winter, and farm people and other interested parties were invited to attend for free or at a nominal cost. Institute programs included sessions on agricultural developments and practices and sometimes Chautauqua-like, general interest presentations as well. Thus they served a recreational as well as an educational function. In the early years of the Station all of the principal members of the staff appeared on institute programs, acquainting local farmers with relevant agricultural research and building good will. Despite its educational and publicity value institute work was rigorous, especially during North Dakota winters, and it took the scientists away from the Station and College for days at a time. When finances permitted, the work was consigned increasingly to professional institute lecturers; eventually agricultural extension replaced the institutes altogether.

The early years of the Station also witnessed some tentative steps toward building support groups for agricultural research. In other states politically astute station scientists—called "research enterpreneurs" by historian Charles Rosenberg—identified likely client groups, often composed of commercial farmers producing particular commodities, and attempted to establish relationships with them. In these relationships client groups benefited from research tailored to their needs while stations benefited from powerful and reliable political support. In North Dakota the task of establishing client groups was made difficult by the lack of commodity-based organization in agriculture and by the pervasiveness of wheat production. The fact that everybody grew wheat meant that it was not the concern of anybody *in particular*—the importance of the crop dictated that the Station would research it whether it had support from an organized client group or not. The Station did have some success in stimulating support groups in other areas. Edwin Ladd used the specter of oleomargarine in his successful effort to spark formation of the North Dakota Dairymen's Association. The early years also saw tentative approaches to flax and livestock producers by Bolley and Shepperd, respectively. Organization of client groups was further advanced around the turn of the century, when John Henry Worst was director, and similar organizations provide important support for the work of the Station today. As is the case in so many other areas the outline of this development can be traced to the Station's early days.[23]

III

The progress of the Station in organizing and popularizing the work was threatened in 1893 by political controversy and economic depression. Of the two problems the political controversy was the more serious. The political fight at the Station began on the state level early in 1893. In the state elections of 1892 an "Independent" ticket headed by Larimore farmer Eli C. D. Shortridge swept all state offices save one. The Independents were the North Dakota version of the Populist party that had grown out of the Farmers' Alliance movement of the late 1880s. The Farmers' Alliance was an organization of agrarian protest, and its complaints about discriminatory railroad charges, dishonest grain grading, and usurous interest rates found a sympathetic audience among the hard-pressed farmers of the Northern Plains. In 1890 the North Dakota Alliance formed the Independent party, which gained backing from the Democratic and Prohibitionist parties, and in 1892 the new party upset the dominant Republicans.[24]

The Independents were understandably anxious to replace Republicans on state boards and commissions, especially those who were allies of Alexander McKenzie, the conservative boss who protected the railroads' interests in the state. Three members of the Board of Directors of the Agricultural College and Experiment Station whose terms were ending—H. F. Miller, James B. Power, and A. M. Tofthagen—were slated for replacement by Shortridge. Unfortunately for Shortridge the Republican-controlled state senate would not confirm his appointments, and the lame ducks refused to vacate their seats. Shortridge challenged the holdovers through legal action, but the state supreme court—which was, probably not coincidentally, predominantly Republican—found that in the absence of confirmed replacements directors whose terms had expired could keep their seats. In anticipation of this finding Shortridge took steps before the court's decision was rendered to remove the entire board, save one friendly member, for cause. It was at this point that the staff became involved.

On the basis of testimony given at a series of private and public hearings the governor charged the sitting board with a number of offenses ranging from incompetence to willful misappropriation of funds for personal profit. Specifically, the directors were charged with shifting money illegally between Station and College accounts, arranging a sweetheart deal with contractors that resulted in the construction of a shoddy and overpriced College building, directing College and Station business to their personal firms without advertising for bids, and having no vouchers to account for over fifteen thousand dollars in expenditures. Stockbridge was the main witness against the board, and his testimony touched on each of these major areas.

From this vantage point it is difficult to judge the degree of corruption of the Board of Directors. Creative accounting was certainly a feature of many early stations, and relatively fat station funds sometimes found their way into lean college budgets. Indeed, when the station and the college were virtually one and the same, as in North Dakota, strict separation of funds was practically impossible. There were obvious problems in the College building, including sloppy construction and inferior materials, but it is hard to assess blame. The fact that the building is still standing indicates that it was fundamentally sound. It is clear that at least some board members profited from their positions. H. F. Miller and S. S. Lyon, for example, saw to it that College and Station funds were deposited in the Fargo banks of which they were officers, and James B. Power and Peter McKenna regularly sold the Station livestock and supplies on which competitive bids had not been solicited. The unvouchered expenditures are a mystery. The board members argued that most of the money went to day laborers who could not conveniently complete vouchers. This is a plausible though not compelling explanation. The fact is that most of the people who held public positions a century ago expected to profit personally from them. Corruption was defined less by the *fact* of personal benefit than by the *degree* of gain and the *extent* to which the gainer misserved the public in the process of gaining. By our standards the board members were corrupt, but by the standards of their time we can only say that the fact of their corruption was not established.

The friction between Stockbridge and the board arose from the variant attitudes toward public service that each represented. The scientists who staffed the new experiment stations had ideas about the purpose of public service that differed from those of traditional officeholders. They were self-conscious professionals who saw government service as a vehicle for advancing agriculture, science, or the state rather than as a means of enriching themselves. This is not to say that they were oblivious to material rewards, or that they lacked ambition, or that they did not lust after the psychic satisfactions that accompanied personal recognition. But the motives that drew them to public employment diverged sharply from those that compelled most of their contemporaries in government. They were part of a new professional class in the American bureaucracy, carrying a conception of public service that traditional officeholders had difficulty understanding.

Perhaps the conflict implicit in these divergent motives could have been avoided had the board not materially impeded the operation of the Station. Beginning in the early months of his tenure Stockbridge struggled with the board over how and where the money should be expended. As a trained, professional scientist, hired to run the Station in the public interest, Stockbridge naturally wanted to control the money. The board, at least partially interested in benefiting from the appropriations, determined to retain fiscal control in its own hands. This created a clear conflict between Stockbridge and his colleagues on one side and the board on the other. In 1895 Ladd wrote to Director A. C. True of the OES, "as you are aware the conditions in the past have been such that we have not been able to expend the money wholly along the lines that the staff would desire to do." To Ladd and the other scientists the problem of corruption was important, but it was secondary to the main difficulty—i.e., that a nonprofessional board rather than a professional staff controlled expenditures.[25]

Board members also involved themselves in personnel matters. In 1892 Power, acting on his own, hired Willet M. Hayes away from the University of Minnesota to be Station agriculturist, giving Hayes's wife Clara the position of home economist at the same time. After his arrival Hayes was authorized by Power to hire two assistants. Stockbridge was outraged by these actions. It was not that Hayes was incompetent or corrupt; he certainly measured up to the standards of the time and was well known as an enthusiast for agricultural improvement. What rankled Stockbridge was that he was not even consulted regarding the hiring. Increasingly, professional expertise was being ignored by the board in a number of areas.

Stockbridge proved to be the ultimate loser in this conflict. Learning of his secret testimony against it, the board dismissed him in May, replacing him with James B. Power. Power had a broad range of experience and some political attributes that made him an attractive successor to Stockbridge from the board's standpoint. For many years he was the land agent for the Northern Pacific Railroad, in which capacity he had done much to popularize the state. It was he, for example, who conceived the "bonanza farm" experiment that brought much favorable attention to the area. He was also a progressive agriculturist who owned and operated an extensive grain and livestock farm called Helendale in southwestern Cass County. A prominent Republican and a leading prohibitionist, he was widely respected throughout the state and enjoyed support from both the progressive and the McKenzie wings of his party. Moreover, he was not part of the Fargo "gang" so prominently represented on the board. Still, Power was not well qualified to run a college and experiment station, even by the standards of his time. He was installed by an embattled board with the hope that he could help save it from the governor. As he wrote to former Governor Andrew Burke in the fall of 1893, "my principle business is that of doing my share of what is necessary to keep the Agricultural College from being absorbed by the Populists." Obviously, Power himself recognized that his educational or scientific credentials had not decreed his selection.[26]

Power did a good job of keeping the Populists out. The governor continued his attempt to remove the board for cause, periodically even resorting to the extreme step of requesting that the United States Treasury withhold Hatch Act funds from it. The board continued to resist, denying that the governor had any authority over it whatsoever. In November of 1893 the state supreme court agreed with the position of the board, and Shortridge was left with no alternative but to drop this acrimonious quarrel.

The controversy ended, but the cost was substantial. Stockbridge, the high-minded professional who built the Station, was gone, and in his place sat a politician. Recognizing the unattractiveness of his position as a principal in the fight, the talented Hayes left in 1893, continuing a career that eventually took him to the USDA as assistant secretary. He was replaced by his brother-in-law, John Shepperd, who eventually served the institution for half a century, including several years as College president. The staff had supported Stockbridge, and the effect of the fight on their morale was difficult to determine. Stockbridge had worked hard to popularize the Station and its work among the farmers of the state. Surely this sordid struggle, with its implications of corruption and political control, did not advance that effort.

Perhaps those interested in the Station derived some cold comfort from the knowledge that the North Dakota facility was not the only one caught up in political controversy. Indeed, political struggles, especially between boards and staffs, were endemic in these early years. As early as 1888 the commissioner of agriculture decried "political interference and manipulation" in the stations, and in 1893 Secretary of Agriculture J. Sterling Morton complained that "some of the moneys appropriated for experiment stations have been diverted from legitimate public purposes and turned to those of a personal and not patriotic character." Three years later Morton turned his attention to the question of professional versus political control: "An experiment station's proper management has no possible relation to any political party whatsoever," he wrote. "It should be in the hands of the best experienced and most practical scientists of the State or Territory."[27]

Their recognition of the ubiquity of political controversy might have been one factor that prevented Stockbridge's colleagues from following him out of North Dakota. Another might have been their appreciation of Power's deep interest in and commitment to the Station and its work. While not a professional educator or scientist, the new president believed in progressive agriculture and in popularizing scientific work, especially through farmers' institutes. Both Power and his son, J. A. ("Allie") Power, maintained for many years a vital interest in the Station and a cordial relationship with the key staff members. Then there was the problem raised by the depression that struck in the spring of 1893 and continued in virulent form for over two years.[28]

The Depression of 1893, second only to the Great Depression of the 1930's in severity, was a nationwide downturn that was especially hard on North Dakota. A young, underdeveloped, undercapitalized state, dependent on a single commodity for its livelihood, North Dakota was particularly vulnerable to national economic setbacks. As tax collections fell toward zero the state could not meet even the minimal obligations to its citizens that it had assumed. When Roger Allin became governor in 1895 state services were virtually suspended for a year. This heroic treatment saved the state from bankruptcy, but the cost was great. For the Station the depression meant that needed state aid would not be secured. On the other hand, the annual Hatch Act appropriation assured that the Station and College would survive and even made them appear prosperous relative to other state institutions. On the national level the depression meant retrenchment. Even the experiment stations, with their federal subsidies, were fortunate to hold the line and were not prepared to offer new positions. Prudent people—such as the scientists at the NDAES—did not quit their jobs during the depression regardless of the distasteful aspects of their employment. Thus, ironically, the depression helped the Station hold its people, preventing what could have been a disastrous exodus in the wake of the Stockbridge affair.

When McKenzie Republican Roger Allin replaced the unhappy Eli Shortridge as governor in 1895, James B. Power wondered what the future held for him. He was willing to stay on, but he thought a professional scientist should be secured to head the institution. At the same time he was a political realist who gauged from Allin's appointees to the board that he would probably be replaced by one of the governor's political allies. He was correct in that perception, because Allin appointed his lieutenant governor, John Henry Worst, to head the College and the Station. This must have been a disappointing development for those who hoped for professional control and the insulation of the Station from politics.[29]

3
BETTER DAYS

The idea that any intelligent citizen can manage the details of a school or college has been so deeply rooted in the popular mind that it can be eradicated only by long and persistent effort.

A. C. True, 1896

To you more than to any other person, not excepting the workers themselves, must belong much of the credit which may have accrued to the people of the state from the success of the investigations of the Experiment Station. Under your kindly supervision it has been possible for each investigator to follow any ideas arising from any investigations which he may have undertaken to a legitimate conclusion, and to bring them before the public in such a manner as to be directly usable by those for whom such work was undertaken.

H. L. Bolley to J. H. Worst, 1913

When John Henry Worst became the third person in less than three years to hold the dual position of College president and Station director at Fargo, the institutions over which he presided had barely started to make an impression on the state. The staff was small, finances were constricted by depression, students were few, experimentation was scarcely under way, and outreach to the state was just beginning. The indifference of most farmers, the disinterest of the community, and the political controversies that had climaxed—but had not ended—with the dismissal of Horace Stockbridge had eroded morale. Lacking broad popular support, but commanding attractive resources, the Station continued to present an inviting target to predatory politicians.

At first glance, the new president and director must have seemed to be just another spoilsman. Forty-four years old in 1895, Worst had come to North Dakota from his native Ohio in 1883. He homesteaded in Emmons County, in the south central part of the state, among others of German and German-American backgrounds. He served as county superintendent of schools and then as a state senator. His loyalty to Republican boss Alexander McKenzie was rewarded in 1894 when he was nominated and elected to the lieutenant governorship. Worst's formal qualifications for leadership of a scientific institution were as unimpressive as Power's had been. He had practical experience as a farmer but had attended college only briefly. At best he seemed just another "intelligent citizen" whose management of educational and scientific institutions Office of Experiment Stations Chief A. C. True found repugnant. At worst he might be a McKenzie henchman lusting after patronage and graft.[1]

Worst proved to be a pleasant surprise in office. As College president he oversaw a building boom and a rapid increase in enrollment and was regarded with affection and loyalty by students and faculty alike. Wise enough to know what he did not know, Station director Worst left the control of experimentation in the hands of the scientists. He perceived his job as popularizing the Station's work, building public support for it, and insulating his subordinates from outside pressures. This job needed to be done for the Station to thrive, and Worst did it well.

It was at least partially due to Worst's efforts that the Station enjoyed what might be called its "silver age." State and federal support grew dramatically, physical facilities were greatly improved, branch stations were added, staff size multiplied, Station activities increased in number, type, and complexity, and research of unquestioned value to the state and the nation was undertaken. By 1914, when a reluctant Worst was half-forced to relinquish the directorship, the Station was an important and respected part of the life of North Dakota. As H. L. Bolley wrote when Worst stepped down, "much of the credit" for what had happened was rightfully his.[2]

Much, but not all. Worst was fortunate to direct the Station during a period of economic well-being and one in which agricultural research benefited from broad public support. The state enjoyed the "Second North Dakota Boom" between 1898 and 1915. Population jumped from 270,000 to 637,000, with the western and north central sections of the state enjoying the bulk of the growth. Agricultural development was spectacular in the fast-growing regions. Between 1898 and 1915 the number of farms in Billings County went from 2 to 632. Ward County went from 377 farms in 1898 to 1,843 in 1915 and the number of farms in Williams County jumped from 67 to 2,342 in the same period. Cultivated acreage rose in proportion to farm numbers. As a group American farmers enjoyed the "golden age of agriculture." General prosperity characterized agriculture, as farm prices rose faster than the general price level and farm incomes swelled. Policymakers' concerns about low agricultural productivity and the high cost of food spurred increased appropriations for agricultural research.[3]

Worst had the good sense to recognize these advantages and pursue them to the fullest. He pressed the legislature for increased appropriations. He made the work popular and necessary for the state through an aggressive outreach program.

He created client groups to support the Station and plead its case in Bismarck. And he encouraged and sheltered an ambitious and courageous program of public-interest oriented research—often regulatory in nature—that popularized the Station in the state and beyond. In capitalizing on the times Worst and the Station staff took risks and made errors that haunted them later, but they created a major institution where a minor one had been before.

I

John Henry Worst is an enigmatic figure in the history of the NDAES. His staff remembered him as a quiet and kindly man who kept his own counsel. Most of those connected with the institution regarded him with respect and even affection, but few really seemed to know him and what motivated him. Understanding him is especially difficult because he left no private papers or letters, and such of his letters as appear in the voluminous Bolley and Shepperd collections are generally brief, formal, and unrevealing.

Worst's public remarks reflected a generally conventional view of North Dakota agriculture, its future, and the relationship between it and the Experiment Station. Like most turn-of-the-century commentators on rural life, he celebrated the small farmer, whom he saw as agriculturally and morally superior to the large-scale producer. "The small farmer . . . ," wrote Worst in 1898, "gives greater diversity to crops, practices a more intensive system of cultivation, and does not rob his farm of its fertility for immediate gain. His farm is his home and the surroundings are beautified with lawns, trees and shrubbery. He expects to . . . bequeath it to his children, more highly improved and more fertile than when he began." Worst was scornful of the bonanza farm, despite its significance in popularizing agriculture in the state. "It is scarcely a home," he wrote, "seldom beautified by tree, shrub or flower, [and] is a system of soil butchery, wherein the fertility is sacrificed for present gain." Because of the large farms and their waste, "the State is kept in a backward condition in point of population and wealth." Celebration of the small and highly productive farm was popular in a state that believed achieving its destiny as another Iowa or Wisconsin depended on development of a dense rural population base. It also echoed those national commentators in the thriving Country Life Movement who saw the ideal farm home as an antidote to rural depopulation and who perceived in a more diverse and intensive agriculture the solution to static productivity and rising food costs.[4]

Worst anticipated a crucial role for the Station in the creation of a prosperous commonwealth of small farmers, predicting that "as the population grows denser and farms are subdivided will scientific agriculture displace present methods, and the work of the Experiment Station become appreciated." The Station would be esteemed because it would unlock the secrets of successful agriculture in a difficult region. "No country in the world of equal fertility requires so much study to master its agricultural resources," he wrote in 1897. "In no country is it more essential that the farmer . . . understand the laws that control plant growth." By discovering and conveying to farmers the scientific laws Worst believed to be essential to success, the Station would turn them into intelligent, productive professionals who maintained soil fertility and produced a diversity of crops efficiently. "Agriculture . . . ," he told the convention of the Association of American Agricultural Colleges and Experiment Stations in 1909, "is destined to become the great profession of the future." And a professional, intelligent, scientifically oriented base of small farmers would mean general prosperity for the state. Thus, "for every dollar expended for a more scientific and intelligent system of agriculture many dollars . . . accrue to the country."[5]

In retrospect, much of Worst's vision seems as unrealistic as it was grandiose. North Dakota could only partially overcome limitations imposed by climate and location. Farmers resisted diversification, and agriculture remains extensive rather than intensive. Moreover, science and professionalization have historically done more to doom than to encourage the small farmer. But it is important to remember that Worst accurately expressed the spirit of optimism and promise that animated agricultural science at the turn of the century. Views such as his were also common in North Dakota, which, like every other young state, saw prosperity and greatness as inalienable parts of its birthright. And he did present an inspiring future to the Station, promising it an indispensable role in helping North Dakota achieve its destiny.

The Station might also help John Henry Worst achieve his destiny. Worst was a successful politician whose ambitions did not necessarily cease when he came to Fargo to run the College and Station. Indeed, his challenge to Asle Gronna for the Republican nomination to the United States Senate in 1914 and his service as commissioner of immigration in Governor Lynn Frazier's Nonpartisan League administration are indications that he continued to cherish high office. From what position could one better advance a political career in an agricultural state than that of director of a successful, popular, and respected experiment station?

When Worst came to the Station in 1895 it did not have the appearance of being a very promising vehicle for the advancement either of the state's agriculture or of his political career. The professional staff consisted of five men. The physical facilities in which they worked were inadequate and sometimes primitive. The Station was committed to encouraging livestock production, but it lacked sufficient housing for animals. Edwin Ladd complained that Chemistry was crowded into quarters "never designed or intended for this kind of work and wholly unfit for such purposes. There is no means of ventilation and no provisions for carrying off noxious fumes and gases . . . I . . . each night . . . find myself exhausted from breathing foul air freighted with obnoxious gasses, and some days even poisonous gases." Worst bemoaned the lack of a sewage system. "The drainage of the main college building, . . ." he wrote, "is pumped out by steam and allowed to freeze in the rear of the building, producing . . . unpleasant and unhealthful effects every spring." No phenomenon that made rational people cherish North Dakota winters and dread the coming of spring can be very attractive.[6]

Limitations in staff and physical facilities were only the most obvious shortcomings of the Station. They were merely symptoms of the larger problem that the Station had as yet made little impact on the state. Certainly, the scientists received and answered inquiries, held institutes, and wrote bulletins, but the Station was not yet an institution playing a major role in the state's agricultural economy, nor was it even an entity of which most North Dakota farmers were conscious.

The Station's unimportance to the general and agricultural publics made it vulnerable to the depredations of self-seeking politicians, and that curse was not quickly lifted. In 1897 veterinarian Theries D. Hinebauch complained to Director True of the OES that the board had dismissed him for no reason and had given his position to another. True complained to Worst, noting that "the state of North Dakota is just beginning to get its experiment station work on a sound basis and any changes which will unsettle its affairs are especially to be deplored at this time." Worst could answer only that he regretted the situation but that the board had made its promise to the new veterinarian before Worst became director.[7]

In 1899 Worst, Bolley, and Ladd all complained to True of heavy local political pressure to oust the popular College and Station secretary, P. W. Farnham. The source of this pressure was apparently John Haggart, then U.S. marshal for North Dakota. Haggart had been instrumental in getting the College and Station located in Fargo, and his interest in the institution had never been totally selfless. The fact that the secretary handled the Station's money might have led Haggart to covet the position for a political crony. Worst wrote True, "I feel that if the interests of this . . . institution can be jeopardized, and its tried and trusted officers made trading stock to consummate political deals, that it will require years to get it back to where we are to-day." Ladd's plea that True "please consider this as confidential for my own head would not be worth the asking if it be known and the political wheel should turn for revenge" indicated that not much had changed since Stockbridge got the ax. For his part, True abhorred the involvement of spoilsmen in the Station but could not interfere in a state matter. Farnham lost the job.[8]

The OES could indicate its displeasure with political shenanigans at experiment stations by cutting off the money supply, but that was an extreme step that alienated congressmen and did little to advance the cause of agricultural research. If the NDAES wanted to escape patronage politics it would have to do so on its own, but how? Unlike agricultural scientists, who often naively argued that politics had no place in research facilities, Worst recognized that the stations were public institutions and that as such they were and probably should be part of politics in both formal and informal ways. The answer to the Station's problem was to strengthen its political position in such a way as to make it less vulnerable to narrow, self-seeking politicians such as John Haggart. This could be done by popularizing the work of the Station in the state and building support for it, engaging in research of demonstrated value to agriculture, and carrying out functions that benefited the general public. Worst emphasized all three of these areas.

II

Popularization of the work was the field in which Worst excelled. An accomplished and successful politician, he understood the farmers of the state and what they expected—or could be induced to expect—from the Station. He had contacts throughout North Dakota in agriculture, business, and politics. He had an intuitive sense of shifting public demands and a desire to satisfy them. And he was a renowned stump speaker—much in demand by campaigning Republican candidates—who could explain virtually anything to anyone and who never treated his auditors with condescension.

Under Worst the Station not only did new things to popularize itself, but it also did old things better. During the preceding administrations the farmers' institutes had been an important means of disseminating information from the Station, but now institute work was pursued much more vigorously. Worst's first official report displayed some disappointment with the state of the institutes. "Farmers' Institutes have been held in a few localities," he noted, "but without the interest or attention they deserve." By 1900 he was able to inform the governor that "Farmers' Institutes have been held at Lidgerwood, Dwight, Gardner, Larimore, Lakota, Blanchard, Enderlin, Park River, Fingal, Wimbledon, Carrington, Harvey, Minnewauken, and Leeds." He added that "at many of the above named places permanent organizations were effected for future regular meetings." By the next year Worst had concluded that "the demand for farmers' institutes is now beyond our facilities to conduct them without a sufficient appropriation by the legislative assembly to organize a paid corps of institute conductors whose entire time can be devoted to such work during the winter months."[9]

The progress of institute work provides a nice case study in the popularization of the Station. Well-conducted institutes carried the work of the Station to the farmers, and successful meetings generated support for similar gatherings in other localities. The organization of local support groups allowed the Station to identify and energize citizens who certainly favored the institutes specifically and probably the Station generally. A burgeoning institute movement, supported by multiplying constituent groups, conveyed potential political clout. Local demand for institutes in North Dakota, stimulated by the Station, led to increases in state funding that eventually allowed the employment of a professional institute staff. A vigorous institute movement also resulted in a favorable public perception of the Station, which now seemed more important to agriculture than ever before. And a visible Station with a favorable public image was less vulnerable to manipulation by gang politicians.

The need for increased visibility stimulated other activities as well. The distribution of bulletins was expanded. H. L. Bolley developed a weed collection, which he sold to rural schools for a nominal fee. The Station vigorously distributed seed, both of experimental crops such as broom corn and durum wheat and of superior varieties of established crops that it had investigated. As the testing program matured, the Station was able to offer to the public more varieties that had been tried thoroughly in the field. Station work was also popularized through a farm newspaper, the *North Dakota Farmer and Sanitary Home*. Founded by Edwin Ladd in 1899, this paper became a virtual house organ for the Station, regularly carrying stories about the scientific work at Fargo along with the more common farm periodical fare. When Worst was director few opportunities to popularize the NDAES were missed.

The Station also continued eagerly to offer advice to virtually anyone who solicited it. The increasing visibility of the Station, the rapid growth of the state, and the special challenges of agriculture in the semi-arid western portion of North Dakota stimulated a flood of inquiries, most of which were answered the next working day. Everybody received many letters, but most farmers wrote to agriculturist John H. Shepperd, whose knowledge of crops, animals, and cultivation practices made his advice especially valuable.

Shepperd replaced his brother-in-law Willet M. Hayes as Station agriculturist in 1893. The twenty-four-year-old Iowa native was in a delicate situation given the acrimonious firing of Horace Stockbridge and the conspicuous position of Hayes in that controversy. But Shepperd was a helpful colleague and an unfailingly kind and considerate man, and he was soon accepted by the other young researchers at the Station. Although he was more conservative and less flamboyant than his more famous colleagues, H. L. Bolley and E. F. Ladd, he was confident in himself and ambitious for advancement. His major goal—to become director—was never realized, but he did serve as president of the College during the troubled thirties.[10]

Shepperd was called on to answer all sorts of questions—sometimes one hundred or more per week—from North Dakotans and from outsiders interested in the state. Most of these sought advice on crops, animals, and methods, but a sampling of Shepperd's turn-of-the-century correspondence indicates the range of subjects on which his opinion was solicited. For example, a Virginia man interested in North Dakota wrote, "I would very much like to find out what part of your state has *cheap lands* . . . of a strong lasting nature . . . . I do not wish to go into debt." He noted he had one thousand dollars to establish himself as a debt-free grandee. A Minnesotan writing in 1901 asked for the winter temperature in Burleigh County and the "exact conditions" for sheep raising there. A New York City woman wrote in 1900 that she had invested in North Dakota land on which she wished to raise wheat. She believed that "'even women'—with the aid of science and the latest approved methods can become successful farmers." The Sisters of Mercy at Belcourt requested information about destroying "French Weed," and the steward at the North Dakota Hospital for the Insane at Jamestown wanted to learn how to care for the institution's cattle and hogs. A Rolla farmer was interested in cropping methods in 1901, but he also complained about the local bank and asked Shepperd to "give me the name of some loan company whear I could borry [*sic*] some money and at what interest." Undoubtedly, Shepperd sometimes felt bewildered by the volume and variety of questions put to him, but their very existence indicated the growing importance of the Station and the increasing faith people held in it.[11]

The Station had made some tentative moves toward organizing support groups even before Worst became director. That endeavor became the centerpiece of his popularization program. Worst's most ambitious creation was the Tri-State Grain and Stock Growers Association, founded in 1898. The main function of this group was to hold a winter convention in Fargo; it did little on the local level. The conventions were well attended, however, and they afforded Worst and his staff an opportunity to parade the work of the Station before a sympathetic audience. Other organizations created by the Station included macaroni wheat, Grimm alfalfa, and flax growers' associations as well as numerous animal breeders' groups. These groups were smaller than the Tri-State Grain and Stock Growers Association, but they were often more viable, and they usually enjoyed close relationships with members of the staff.

The organization of supportive client groups was common practice in the early stations. Worst's enthusiastic participation in this endeavor made him the North Dakota equivalent of such directors as Eugene Davenport of Illinois and W. A. Henry of Wisconsin, "research entrepreneurs" who created constituent groups that could be used to generate political power and support. The development of dependable political support was a significant reason for creating client groups, but this practice also allowed interested farmers to have a close relationship with station researchers working on problems of particular concern to them. There were potential pitfalls in this endeavor. Stations sometimes became the captives of the very support groups they had created and found themselves the tools for realizing political and economic agendas developed by others. It was also the case that the inclusion of organized farmers in close relationships with stations often effectively meant the exclusion of unorganized farmers. The interests of general farmers uncommitted to single commodities and of relatively noncommercial farmers tended to be slighted when stations became close to specialized client groups.[12]

Some of the Station's other efforts at popularization also carried risks, though they were not readily apparent at the time. Under Worst the Station established cordial relations with the railroads. He reported in 1900 that the Great Northern, "through the public spirited generosity of President James J.

Hill," had provided free transportation for over two thousand farmers who wished to visit the Station. In 1907 he indicated that the Great Northern and the Northern Pacific had funded the operation of six model farms on their lines, and he thanked those roads, along with the "Soo," for providing "seed trains" on which Station personnel traveled throughout the state lecturing about "seed treatment for the prevention of smut and other diseases, seed selection, seed testing, . . . value of crop rotation and leguminous crops, thorough cultivation, etc." The carriers also provided discount fares to farmers attending the Tri-State Grain and Stock Growers Association convention. This arrangement was mutually beneficial because the roads and the Station were both interested in the growth and prosperity of the state and they generally agreed regarding the main problems of agriculture. On the other hand, the Station's association with the railroads might have besmirched it in the eyes of some farmers. Moreover, when the Station and the carriers seemed to diverge regarding the core problems of North Dakota agriculture the latter did not cease their involvement with the former. Instead, they tried to take it over and purge it. In the real world there are tradeoffs, and the long-term consequences of actions and decisions are not always immediately apparent.[13]

The people at Fargo were remarkably successful in popularizing the Station when Worst was director. But popularization involved more than just style or method. The Station had to prove itself with quality work that responded to the problems of North Dakota agriculture. It did so during Worst's directorship, revealing what proved to be the first in a long series of significant products of Station research.

III

There was no change in the basic direction of Station research when Worst was director. The NDAES remained committed to diversification, especially through the introduction of dairy cattle and other livestock. Everybody at the Station seemed to agree that dairy farming was to be the salvation of North Dakota agriculture and that the NDAES was to lead the way to it. In 1893 dairy specialist E. E. Kaufman was hired. Kaufman was an enthusiastic evangelist for dairy, contending in 1896 that "exclusive grain growing has never been productive of a wealthy and intelligent class of farmers." Writing admiringly of dairy farming in New York and Wisconsin, Kaufman concluded that "North Dakota possesses advantages for dairying equal, if not superior" to those of any other state.[14]

If energy and publicity could have made it so, North Dakota would have become a dairy state. John Shepperd organized the New Salem Breeding Circuit, west of Bismarck, to increase the numbers of purebred Holsteins. Kaufman served as state dairy commissioner and as executive secretary of the North Dakota

Dairymen's Association. Bulletins were produced about such topics as silo construction and the organization of cooperative creameries and cow-testing associations.

Their appeared to be some progress, though the evidence for it was deceptive. The state commissioner of agriculture estimated that milk production had increased from 7,380,000 pounds in 1895 to 17,225,000 pounds a decade later. That latter figure is less impressive when compared to a 1905 wheat crop of over 73,000,000 bushels. The fact is that in 1905 there were fewer than 123,000 dairy cows on the state's 52,000 farms.[15]

Farmers were simply reluctant to share the Station's enthusiasm for dairy farming, and they had good reason to be cautious. Proponents of the milk cow, recognizing the prosperity of dairy farmers in other areas, assumed that the type of farming made the farmers prosperous. What Kaufman, Shepperd, Worst, and other would-be reformers of North Dakota agriculture failed to understand was that New York and Wisconsin dairy farmers were prosperous because of their proximity to huge markets for fluid milk. No such markets existed on the Northern Plains. Because they recognized this, or because they were conservative, or because of the work involved in dairy farming, North Dakota farmers remained wedded to wheat, which continued to be the source of the bulk of farm income. Even those who experimented with dairy cattle and other livestock shifted when the prospects for wheat brightened. L. R. Waldron noted ruefully in 1912 that "one or two good years in succession will change very considerably the farming methods of the area . . . . The corn acreage decreases, livestock is disposed of and the energy of the farmers is devoted largely to wheat." In the final analysis, the dairy promotion program of the NDAES must be seen mainly as an example of the limitations on the ability of an experiment station to alter agricultural practices.[16]

The betterment of rural life continued to be emphasized at the Station during the Worst directorship. The early years of the twentieth century witnessed the flourishing of the Country Life Movement, a multifaceted effort to improve farming and enhance rural living. The theme of the Country Life Movement that rural homes must be made more beautiful and convenient was echoed at the NDAES, especially by C. B. Waldron. Waldron stressed the utility of a beautiful physical environment for shaping people of quality. "Attractive homes constitute an active and potent force in shaping and determining the character of society and its members and after all there is nothing more important than that," he wrote in a 1904 bulletin. "To rear a generation amid surroundings that because of their dreariness and monotony depress and stultify the mind and imagination instead of quickening and enlivening it, is neither wise nor patriotic." Waldron urged beautification on the additional grounds of state patriotism, arguing that "the first step in disarming the critics of North Dakota who berate us for our barren and cheerless homes, is the planting of trees." His experiments with "timber belts" as aids in soil conservation, begun at this time, bore fruit in the shelter belt program of the thirties.[17]

Waldron was not the only Station researcher to explore means of improving rural homes. In 1913 the Station published a bulletin on the use of windmills to generate electrical power for farm use. In his foreward, Worst borrowed a popular argument from the Country Life Movement when he contended that "the rural home must be made a desirable place to live and work, and also it must be invested with things interesting as well as useful and labor-saving, if country boys and girls are to be kept on the farm."[18]

Most research during Worst's directorship was oriented toward the economic side of farm life, and none was more important to the state than that of H. L. Bolley. Like the other pioneer scientists at the NDAES, Bolley was a peripatetic researcher whose interests extended to such diverse areas as weed control, the eradication of typhoid fever, and the use of denatured alcohol for fuel. This broad focus was the result of necessity in an institution with a limited staff and also of choice by a man with an active mind and energetic personality. Bolley's main scientific interest was in crop diseases, and during the first years of the century his early work in that area came to fruition.

Bolley's major contribution to science and to North Dakota agriculture was his discovery of the cause of flax wilt. Flax was an appropriate and potentially valuable alternative to wheat for North Dakota farmers, but the track record of the crop in the United States was not encouraging. After a few good crops in an area the soil would become "flax sick," the plants would suffer from wilt, and production would decline. The common belief of most farmers and scientists was that flax took some needed but unknown nutrient from the soil, contributing to its own demise. Bolley was interested in this problem, and with the help and encouragement of Shepperd, he developed a piece of flax-sick soil by successive yearly plantings of the crop on it. In 1901 Bolley announced that his investigation of the soil indicated that flax wilt was caused by a fungus introduced into the earth by the flax plant itself. Once introduced the fungus carried over in the soil from year to year. This discovery pointed the way toward crop rotation and careful selection and treatment of seed as means of saving the crop. Moreover, the existence of the flax-sick "plot 30" provided a way to develop and test wilt-resistant strains of flax, a task on which Bolley concentrated for much of the rest of his career. Bolley deserves much of the credit for the dramatic 250 percent increase in North Dakota flax production between 1895 and 1905 and for the rise of the state to national dominance in that crop.[19]

Bolley's flax research drew favorable national attention. In 1903 the USDA sent him on an extensive tour of Europe to study the crop there. By 1911 the Flax Development Committee, an industry group created by the trade association of the paint and varnish manufacturers, was providing Bolley with liberal annual subsidies. In 1930 the Flax Development Committee sent him on an investigative trip to Latin America, and it continued to fund him until he retired from active research around 1940. Bolley's flax work also earned him wide acclaim and no little affection in North Dakota. An old friend reminisced in 1941 about a flax-inspection trip he and Bolley had taken in the Little Missouri country in 1911: "There were the stops for repairing tire tubes and the stops for a sandwich or a drink where some farmer would discover that a man named Bolley was in town and it would be but a few minutes before Bolley had an audience which kept increasing with the passing minutes." Bolley's friend concluded by informing him that "at one of those towns I had picked up a lariat and had put it in the back with the intention, when conversations got too long, of roping and throwing you, bundling you into the car, and going on our way."[20]

Bolley's flax work was only the most spectacular of several important lines of inquiry during this very productive period of his career. He continued to work on potato scab and wheat diseases. He refined some of his earlier work on the formaldehyde treatment for wheat seed. And his careful observations of the barberry shrub confirmed the developing suspicion among scientists that it served as a host for wheat stem rust. Much of Bolley's work focused on seed quality. A Darwinian by training and conviction, he firmly believed that good crops produced good seed which in turn produced good crops. Whether he was talking about wheat, flax, potatoes, or any other crop, Bolley emphasized careful seed selection and treatment. His preoccupation with good seed was one of the factors that led him to work for a state seed inspection law. Another was his correct belief that many troublesome weeds entered the state as impurities in grain or flax seed.

After encouraging farmers and seed merchants for several years to send samples to the Station for analysis, Bolley succeeded in getting the legislature to pass a seed inspection law in 1909. Under this legislation the state seed commissioner was empowered to analyze seed for purity and germination and certify it. The process was a voluntary one, but the unspoken assumption was that farmers would soon shun seed that had not been tested and certified. The law also prohibited the sale of seed containing impurities in the form of certain injurious weeds, and it provided for field inspection to help prevent such adulteration. As the main proponent of the law and the premier botanist in the state Bolley was made seed commissioner.

Seed inspection dramatically increased the demands on the already busy Bolley, because farmers and seedmen responded to the law in large numbers. Much of the actual work of seed inspection was delegated to O. A. Stevens, a young graduate of the Kansas Agricultural College who pursued a distinguished career in botany at the Station and who was associated with the College from 1909 until his death in 1979. Field inspectors were also hired, but Bolley did some of the work and was responsible for all of it. In addition to burdening him further, seed work also carried Bolley into treacherous political waters. He recognized that his was a sensitive position and he urged field inspectors to be circumspect in their behavior. He admonished one inspector to "be careful to leave the impression with them [seed dealers] not that we are trying to make things hard for them, but rather to help them to improve the quality of seeds which they place on the market." Still, Bolley could not avoid controversy given the nature of the position, his own outspoken opinions, and his commitment to good seed. He raised the seed dealers against him when he suggested that pure seed growers form a cooperative, and he made powerful enemies when he identified seed houses whose products had been rejected. The farmers appreciated what he did, but his actions carried risks.[21]

Agriculturist John Shepperd was another staff member whose major work was accomplished during Worst's directorship. The position of agriculturist was the most demanding one at most of the early stations. His major job was supervision of the station farm. In the capacity of supervisor he was required to be an executive rather than a skilled scientist, though he did have to execute field experiments in as controlled a manner as possible. He managed numerous plots, buildings, animals, and pieces of machinery, and he directed a large body of farm workers. He also had teaching responsibilities, conducted farm tours, wrote bulletins, answered farmers' inquiries, and participated in institute work.

Like most good agriculturists, Shepperd was a strong executive, though his scientific abilities were indicated by his receipt of a gold medal at the Paris Exposition of 1900 for wheat selection and breeding. Worst demonstrated great confidence in him, making him vice-director of the Station and placing the branch stations under his supervision as they were acquired. In addition to the manifold duties that came with the position Shepperd voluntarily assumed other tasks, such as judging stock at county fairs, conducting stock judging tours to major exhibitions, and organizing breeders' and growers' groups.[22]

Shepperd was an ambitious and energetic man who welcomed responsibility, but he was frustrated by some aspects of his position. His inability to keep competent, well-trained subordinates, which he attributed mainly to the low salaries he could offer, especially irritated him. "I am . . . dismayed with the regularity with which I lose men here at the central station," he wrote one USDA official in 1904. The demand for trained people was high, and in a booming state even common farm labor was hard to retain. The superintendent of the substation at Edgeley complained to Shepperd that he had to fire his first helper because he had "a chronic grievance against common humanity in general . . . 'God Damning' and calling dirty names anyone whom he sees fit to." His next man was "a ball player" who had "to be let off when a ball game is in sight," but the superintendent saw no immediate alternative to him. The frustrations of his position led Shepperd to consider other jobs. Angered by low appropriations and what he considered to be a narrow, ignorant, and unappreciative board, he considered taking the position as director of the Tennessee station in 1904. His colleagues urged him to stay. Ladd wrote that no "sensible man in the state who can appreciate the work and its difficulties wants to see you leave." He added that "your going would be a set back for the College . . . . Your fund of knowledge of things in the state is a great capital for future good."[23]

Ladd's appraisal of his colleague's value to the state was accurate. Shepperd's major contribution to North Dakota, and to the West in general, was his testing of the Campbell System of dry land farming. The Campbell System, which proponents claimed was a means of producing heavy and reliable yields of small grains on semi-arid lands, was developed by South Dakota farmer and promoter Hardy Webster Campbell. The key to the Campbell System was the maximum utilization of soil moisture. It featured packing of sub-surface soil to conserve moisture, quick cultivation following rains to enhance capillary action, and retention of a dust mulch to retard surface evaporation. Compounded of sound scientific principles and Campbell's own untutored observations, this system was embraced by many people around the turn of the century who saw it as the key to growth and prosperity in the West. The Dry Farming Congress, an aggregation of Western railroaders, realtors, bankers, and merchants, boosted the Campbell System, as did most Western state legislatures, governors, and departments of agriculture and immigration. Displaying the wishful thinking endemic in the West, the economic and political elite perceived in the Campbell System the means of making the desert bloom and, not coincidentally, their careers and businesses boom.[24]

There were two main problems with the Campbell System. The first was that it demanded expensive equipment and much labor. The second was that it was untested by reputable scientists. Shepperd believed it was his responsibility to test the validity of the system, and he initiated experiments to that end in the mid-nineties. Subjecting the Campbell System to scientific scrutiny was risky business because believing in it was a matter of regional patriotism, rather as believing "rain follows the plow" had been a few years before. In order to encourage North Dakota farmers to try the system, the Northern Pacific Railroad initiated a program in 1895 of providing Campbell machinery to selected farmers free of charge and of offering cash prizes for the largest yields obtained through Campbell methods. A serious investigator might find that the Campbell System did not work as it was supposed to, and nobody could welcome the task of carrying that bad news to the powerful boosters of the method.

But that was just what Shepperd and his assistants did. Based in skepticism about the claims of the boosters of the Campbell System, their experiments first suggested and then eventually showed that while the method was expensive and labor-intensive it did not deliver the promised benefits. They concluded their experiments in 1901 with the judgment that "the Campbell method has yielded small crops of wheat and has given no appreciable saving of soil moisture," and they urged farmers to stick to traditional practices. Whether Shepperd came under the pressures those in other states were subjected to when they tested the Campbell System was never revealed. What is clear is that he was proud of saving North Dakota farmers from unrealistic dreams and from the pointless expenditure of thousands of dollars for machinery and labor.[25]

The solution Shepperd and his assistants favored for the problem of agriculture in the semi-arid portions of the state was diversification, with sound rotation of crops and a heavy reliance on livestock. He deplored the wheat monoculture everywhere, but he thought it was especially unwise in western North Dakota and he was not reluctant to say so. When Shepperd advised a questioner against wheat farming in the southwestern portion of the state an angry legislator wrote Worst that the agriculturist "has caused a great deal of adverse feeling for the Agricultural College and its management in our part of the country." Shepperd replied to this "caustic letter" by defending his advice and asserting that he would give it again. There was no question John Shepperd had the courage of his convictions.[26]

Shepperd championed livestock, but his efforts to get North Dakota farmers to share his enthusiasm resulted in much frustration for him. He was especially irritated with what he perceived as an aversion to stock raising among Norwegians, the largest ethnic group in the state. "I think the Scandinavians have worked with fish too much," he told a bankers' convention in 1911. "They do not understand anything with warm blood in it." In view of the reluctance of farmers to raise animals, the development of alternative crops to hard spring wheat became ever more important to the agriculturist. In that context Shepperd worked intensively with corn, potatoes, and other crops. He was also an enthusiastic promoter of durum wheat, which became the center of another controversy in which the NDAES was involved.[27]

When durum wheat was introduced into the United States by the USDA in the 1890s it had two things to recommend it. First, as its popular name "macaroni" wheat indicated, it was the basis for a superior semolina, the type of flour used to make pasta. And second, it was a variety that returned impressive yields in semi-arid areas. The later discovery that it enjoyed remarkable resistance to stem rust added to its appeal. Shepperd, who was always on the lookout for marketable alternatives to the hard bread wheats that dominated the state's agriculture and for crops that were appropriate for the drier parts of the state, immediately showed an interest in durum. By the late nineties he was distributing hundreds of samples of the grain to interested farmers and was propagandizing for it among millers and bakers in the state.

As a result of the efforts of Shepperd and others there was a steady expansion of acreage in durum, but the future of the crop was clouded by two factors. First, there was insufficient demand for pasta in the United States to justify large-scale durum production. Second, millers of bread flour did not want durum because they argued it produced less flour per bushel than hard red spring wheat, required different milling machinery, and produced a yellow flour that bakers and consumers shunned. The millers' attitude became an especial problem in 1903 and 1904. Because of its resistance to rust and its vigorous promotion by the USDA, durum production leaped forward, with most of the increase coming in the Dakotas. Durum promotors recognized the millers' bias against the crop, but they believed, with the USDA's chief cerealist Mark Carleton, that "the millers must see that the wheat has come to stay and that they might just as well begin now as any time to use it in their mills." Carleton's optimism was not immediately justified, because the huge crops sold for twenty-five to thirty cents less per bushel than hard red spring wheat when buyers could be found. Growers were understandably eager to find out who was responsible for this lamentable situation. Noting that local dealers paid thirty cents less for durum than for hard red spring, a Kenmare farmer wrote the Station that "owing to the ridiculous low price offered . . . for durum wheat, I assume the authority to write you on behalf of the durum producers of this immediate vicinity, in hopes of learning the best markets for this variety of wheat." The Station could not provide a satisfactory answer.[28]

The durum promoters at the USDA, the NDAES, and elsewhere blamed the millers for failing to take the wheat or discounting it so severely. The millers, whose major trade publication stigmatized durum as "goose wheat" and a "bastard variety," attacked the USDA for first over-promoting it and for then blaming the milling industry for its low price. *The Northwestern Miller* sputtered that "the agricultural department apparently proposes to force [durum] into consumption regardless of the laws of supply and demand and in defiance of the opinion of thoughtful wheat raisers and flour millers." A week later the *Miller* editorialized that "the official advocates of macaroni wheat have already done great harm to the northwest; they have misled its farmers; maligned its millers; sown the seeds of class hatred and prejudice, . . . and have generally exercised their official power and influence to dangerous and mischievous ends." The durum fight was joined, and the redoubtable Edwin Ladd soon entered the fray on behalf of the scorned variety.[29]

Ladd was another of the early scientists whose energy was a marvel. He and his growing staff were responsible for all of the Station's chemical analysis—of crops, fertilizers, weed killers, formaldehyde, commercial animal feed, and insecticides. He analyzed samples of soil, water, food, drugs, liquor, agricultural products, and anything else North Dakotans sent to the Station. After 1900 he became the untitled state chemist of North Dakota and assumed official responsibility for analyzing food and drug products, paints, and a variety of other items. His chemical work alone was so demanding and comprehensive that by 1912 he had broken his department into fourteen functional divisions. In addition to his work in chemistry, Ladd tirelessly promoted sugar beets, edited the *North Dakota Farmer and Sanitary Home*, taught classes, wrote bulletins, and did institute work. In his spare time he served as Station meteorologist, assuming the unenviable task of attempting to discern clear patterns in North Dakota's weather. None of these duties was performed in a perfunctory manner. Ladd was interested enough in weather, for example, to request funds from the OES that would enable him to test his theory that hailstorms could be dispersed by firing artillery into them. Director True politely turned him down.[30]

Ladd became involved in the durum wheat controversy in 1905. At that time the legislature appropriated funds for the erection of experimental milling and baking facilities at the Station. Through operation of these by trained technicians the state could receive definitive answers to questions involving effective flour yield of various wheats, relative difficulty of milling different varieties, baking performance, gluten content, and nutritional qualities. Ladd's milling and baking studies exposed abuses and developed empirical evidence that contributed to passage of the Grain Standards Act of 1916, federalizing grain grading. His studies also allowed him vigorously to champion durum.

Some of Ladd's research conclusions could not have been very encouraging to North Dakota farmers. In 1910, for example, he announced that the quality of flour produced with winter wheats from Kansas and Nebraska was superior to that derived from North Dakota winter wheat. He also concluded that the wide variations in quality within each wheat classification made the whole grading system suspect. That was good news for those producing lower grades, but it was not heartening to farmers selling No. 1 Hard. On the durum issue Ladd's conclusions were more gratifying. He argued that the differences in flour yield and milling difficulty between hard red spring wheat and durum justified no more than a five-cent-per-bushel price differential rather than the prevailing twenty to thirty-cent gap. Why was the differential so great? Ladd believed it was due to a number of factors, including the millers' anti-durum bias and "the possibility that the elevators have charged a larger margin of profits in handling this class of wheats than for fife and blue-stem." The milling industry, which traditionally scoffed at "the trivial demonstrations of an amateur miller in an experiment station," was unimpressed by Ladd's conclusions. But Ladd's milling and baking experiments formed only one prong of his offensive on the durum issue. The second part of his attack involved bleaching.[31]

Bleaching was a major innovation in milling in the early twentieth century. In their natural states no flours are as white as bleached flours. Hard spring wheats are milled into flours of a cream or ivory color, winter wheat flours run from dark ivory to light yellow, and durums produce darker yellow flours. Other grains produce still darker flours, and flour made from rye is virtually black. Bakers and consumers traditionally preferred lighter flours, which is a major reason why "Minnesota Patent" flour made from hard red spring wheat commanded a premium price in the late-nineteenth century. Through the use of a chemical process introduced in this country in 1904 flour could be "bleached," effectively removing most of its color. The millers who engaged in this expensive process argued that bleaching purified the flour, but in fact they bleached the product in order to make it more aesthetically pleasing to consumers.

Bleaching was a potential threat to North Dakota wheat producers in two different and somewhat contradictory ways. First, by neutralizing the natural advantage of hard red spring wheats, it threatened to diminish the price premium traditionally enjoyed by those varieties. Second, it potentially allowed millers to reap an unfair profit from durum, which they could buy at a low price, bleach, mix with flours ground from traditional bread wheats, and sell at a premium price to an unsuspecting public. If millers were using the bleaching process to conceal durum in more conventional bread flours they were cheating both the producers and the public. Ladd believed that they were doing this, and he determined to use his authority as state food commissioner to stop them.

In 1903 the legislature had added to Ladd's many duties—at his own urging—by making him state food commissioner and giving him the power to remove from the market foods adulterated by harmful additives. In 1906 Ladd's chemical and baking experiments with bleached flour showed "the presense of nitrous acid in the flour" which gave "the bread a decided odor and taste." He added that "it would seem that the amount present in the bread would be harmful physiologically." Putting on his food commissioner's hat, Ladd declared in 1907 that bleached flour was adulterated and banned its sale in the state. The millers defended bleaching by arguing that the process merely purified quality flour and aided in its preservation, and they denied that any of their number would be able or willing to fool the public by bleaching lower grades and mixing them with higher ones. They also enlisted a complaint scientist, Harry Snyder of the Minnesota Agricultural Experiment Station, to argue that the level of nitrogen peroxide in bleached flour was too minute to harm anyone.[32]

Ladd's statewide ban on bleached flour went to court, and early in 1909 Fargo Judge Charles Pollock found, rather paradoxically, that while nitrogen peroxide did not exist in harmful quantities in bleached flour, it was an adulterant and Ladd was justified in his action. The *Northwestern Miller* was disappointed, and it castigated Ladd as a "professor . . . of . . . peanut politics." Worse was yet to come for the millers. Later in 1909 their old nemesis the USDA agreed that bleached flour was adulterated and prohibited its sale everywhere in the country. That ruling stood until 1919, when the government agreed to allow the millers to sell bleached flour if they clearly labeled it as such. The effect of bleaching remains something of a controversial issue, but today most scientists agree that bleached flour is not harmful to humans.[33]

The effects of all these activities are not easy to judge. Durum prices rose before and during World War I, but it is impossible to determine the degree to which this was due to the efforts of the NDAES and other promoters. A waxing European market was certainly part of the reason for rising prices. It is clear that the researchers at the Station showed themselves to be loyal and resourceful promoters of what they understood to be the interests of North Dakota. And their efforts on behalf of the crop, in combination with the natural advantages of the "Durum Triangle" in the northeastern and north central part of the state, helped make North Dakota the leading durum producer in the country. From another perspective the durum controversy raised serious questions about the ability of science to solve public problems in a conclusive manner. The behavior of Ladd and Snyder must have disillusioned those who wanted science to be a neutral and value-free force for public good.

IV

The durum wheat controversy illustrated the way in which producer-oriented research blended with public-interest research when Worst was director. Sometimes it was difficult to separate the two. In a state in which 80 percent of the people were on farms and in which most of the farm income came from wheat, one could almost conclude that anything done for wheat and its producers served the public interest. L. R. Waldron discovered in 1903 that in forth-three of forty-seven samples he tested, elevators had docked grain excessively, costing "the farmers of the state . . . about 400,000 bushels of grain in one season." In a state in which everyone's prosperity was tied to one crop, Waldron's research findings were emphatically in the public interest. Willet Hays had equated support for spring wheat with state patriotism during the 1890s, and he did not much overstate the case.[34]

During Worst's directorship the NDAES became nationally famous for public-interest research involving consumer protection. The burden of undertaking this work fell to the Station in part because it was the main locus of scientific expertise in the state, but this factor should not be overstated. It is clear that key figures at the NDAES—such as Worst, Ladd, and Bolley—were committed to scientific work that would benefit a broad public. In this commitment they reflected the common assumption of many turn-of-the-century scientists that the educated professional had a special responsibility to use his or her knowledge to serve the community. Their commitment also mirrored two themes of the era of progressive political reform in which they lived:that science could provide answers to many public problems and that there was a community interest that should be upheld against the assaults of selfish individuals. It is also likely that Worst recognized the utility of public-interest research in making the Station an important and vital factor in the life of the state, rendering it stronger, more visible, and less vulnerable politically.

Most of the consumer protection work at the NDAES was undertaken by Ladd, who gained national attention through his championship of pure food and drug products. The early twentieth century was a period of growing concern about the quality of food products consumed by Americans. Public interest in the nutritional properties of food spurred an ambitious and fruitful research program at the USDA and a number of experiment stations. The related concern about the presence of potentially harmful additives in the processed foods that Americans increasingly consumed was reflected in demands for the investigation and enforcement of food purity. Ladd had been interested in this latter area since his early days at the Station, and the legislature passed a pure food law at his urging in 1901. This law proved deficient because, while it banned adulterated foods from the state, it neither created the means whereby food products were to be inspected nor gave State's attorneys the authority to prosecute violations. Ladd was not reluctant to point out the defects of what was in effect merely an admonitory statute. "It is little less than criminal that food products should be allowed to be sold in the state containing such large quantities of chemical preservatives and coal tar dyes, as have been found in some of the canned goods, catsups, and preserves," he wrote in 1902. He also complained that it was impossible to punish those who preserved milk by adding formaldehyde to it. "Milk containing formaldehyde certainly can not be a safe food for infants and invalids," he reminded his readers. Ladd suggested that the law be strengthened by lodging the power of inspection in the Station and allowing State's attorneys to prosecute violations. In 1903 the legislature complied, passing with little change a pure food law written by Ladd and making him state food commissioner.[35]

Food Commissioner Ladd became a virtual folk hero in North Dakota. Food processors and manufacturers from outside the state pursued the strategy of suing him—and sometimes Worst as well—for libel when he declared their products to be adulterated, which he always did publicly. Rather than intimidating Ladd these suits made him seem a sort of North Dakota David battling for his people against out-of-state Goliaths. In a small and isolated state that often felt oppressed by outside capital, this was a perfect image to have. Ladd never lost a suit, and the legislature, in which he enjoyed almost total support from all factions of both parties, never failed to vote funds to cover his legal fees. His defense of the stomachs of North Dakotans won their hearts, and eventually won him a seat in the United States Senate. The lawsuits against Ladd, his bristling self-righteousness in pure food matters, and his public naming of names obscured the conservatism of much of what he did as food commissioner. Only in the most extreme cases would he take retailers to court and he always preferred persuasion to prosecution. Ladd indicated in 1906 that his major goals as food commissioner were to "induce manufacturers and jobbers" to obey the law, "give wide publicity to all food frauds," "induce dealers to leave illegal goods alone," and "induce the public to demand pure food products." Reform, not revolution, was Ladd's goal, and he put a great stress on voluntarism. These were characteristics he shared with most other progressive reformers.[36]

The legislature extended the regulatory responsibilities of the Station in 1905 by ordering it to test paints as a means of uncovering adulteration and preventing deceptive sales. Ladd was given this job, and he entered into it with characteristic zeal. The paint manufacturers had learned by watching the food processors, and they dealt with Ladd in a more enlightened fashion. Instead of suing Ladd, they provided him with free paint samples and with money to advance his experiments. They required in return that he refrain from his usual practice of identifying poor or adulterated products by brand name. The result of this arrangement was that Ladd's experiments were liberally funded, the paint producers received the evidence of the relative performance of various chemical compositions of their products, and North Dakota received better paint. At least, that was what Ladd believed when he argued in 1909 that because of "the enforcement of the Paint Law . . . to-day probably no state in the Union has as high grade paints on its markets as are to be found in North Dakota."[37]

With Edwin Ladd in the vanguard the Station undertook an ambitious program of regulatory work to protect the public interest. Other stations did similar work, but they usually had it forced on them, and they lacked enthusiasm for it. The Office of Experiment Stations also took a dim view of stations exercising police powers. The OES did not consider regulatory work—such as Ladd did with food and paints and Bolley did with seeds—to be an appropriate function for scientific institutions, and it refused to allow the expenditure of federal funds for such purposes. On the other hand, by responding to the needs of a broadly conceived public constituency the Station made itself important to North Dakota. It is difficult to argue with Elwyn Robinson's judgment that, because of public-interest work, "the North Dakota Agricultural Experiment Station probably exerted a more profound influence than did any other such station in the United States." The NDAES was a significant factor in the life of the state, and public-interest research was a large part of the reason.[38]

V

The NDAES flourished under the direction of John Henry Worst. By virtually every measure it enjoyed a level of growth unequaled during any comparable period in its history. Worst's first report showed a staff of five. By 1914 there were thirty-eight on the Station roster at Fargo alone and seven more elsewhere in the state. Facilities were expanded as well, though never fast enough for an energetic band of scientists that could always see much more to do.

One major addition to the Station that enhanced its ability to serve the state was substations.* In 1901 the citizens of Edgeley donated a quarter section of land near town for use as a substation. This gave the NDAES a Drift Prairie location on which to conduct experiments and field tests, and a $250 legislative appropriation allowed work to commence there in 1902. The city of Dickinson and Stark County offered another parcel of land in 1905, which allowed the erection of a substation in the semi-arid country west of the Missouri River. L. R. Waldron was made superintendent of that facility. Substations brought scientific and political benefits, but Worst was dubious about them because he believed they contributed to the tendency to spread resources too thinly. He regretted privately to an OES official that the state had accepted the Edgeley donation without providing sufficient funding for it. "Instead of furnishing us with armor and equipment for our lusty youth they gave us another baby," he complained. Worst

**Until about 1940 these experimental facilities apart from the main station were usually referred to as "substations." From then until 1989, when they were designated "research centers," they were usually called "branch stations." In this volume I will refer to them in accordance with the practice common at the time.*

preferred demonstration farms—on which cooperative farmers would carry out operations in accordance with Station instructions—to substations, and he hoped one could be placed in every county. Most scientists agreed that demonstration farms were of little value, but they were cheaper than substations, and they allowed the Station to show the flag all over the state. By the time Worst was replaced as director there were nearly forty demonstration farms in operation, and more new substations had been organized at Langdon, Williston, and Hettinger.[39]

The growth of the Station was impressive, but expansion was partially fueled by developments over which Worst and his staff had no control. The state was booming, more than doubling its population during the time of Worst's directorship. Moreover, growth was especially rapid in parts of the state where agricultural conditions were particularly challenging. Special needs and high prosperity fortuitously conjoined to increase dramatically state support for the Station. This was also a period of rising federal support for agricultural experimentation. Concerned about low agricultural productivity and seeing science as one solution, Congress passed the Adams Act in 1906. This piece of legislation was designed to free station scientists from classroom duties through the provision of special funds to support research. NDAES researchers did not abandon the classroom, but the Adams Act benefited their work by doubling total federal appropriations.[40]

These special circumstances meant that Worst operated in a no-lose situation. It is hard to imagine that the Station would have failed to grow under anybody's direction during the early twentieth century. At the same time, Worst must be credited with seeing the available opportunities and pressing to maximize them. He was especially aggressive with the legislature. Worst and his staff peppered annual reports with demands for better facilities, more help, and higher salaries. He reminded legislators of the special ability of the Station to address the needs of new settlers and its importance to the future prosperity of the state. And he did not hesitate to upbraid legislators when he thought they had provided inadequately for the Station. In 1913, for example, he complained that the legislature had appropriated only thirty-five thousand dollars to advance dairy work: "The sum provided is too small to make a creditable showing or to meet the demands of an agricultural state in which the dairy industry should be encouraged to the utmost, . . . yet a start can be made with the meagre sum appropriated." Perhaps Worst's own governmental experience convinced him of the importance of prodding the usually cautious legislature. He clearly saw a wide field of activity for the Station, and he wanted state appropriations to reflect its importance.[41]

Worst should also be credited with providing a positive environment for the creative people on his staff. As Henry Bolley's remarks to him in 1913 indicate, Worst encouraged the researchers in their work and defended them from criticism for what they did. Not a scientist himself, he did not attempt to direct them in any meaningful sense of the word. Instead, he gave them their heads, allowing them to do pretty much as they pleased. With this freedom, scientific generalists Bolley, Ladd, and Shepperd enjoyed the most productive phases of their careers, drawing the respect of the state and, at times, the attention of interested observers outside of North Dakota.

Under Worst's benevolent regime the leading figures at the NDAES were able to live their rather romantic notions of what scientists should do and be. They were able to behave as free individuals, guided solely by a desire to use their knowledge to serve farmers and the broader public. They illustrated how valuable they could be to the people as they saved crops, exploded erroneous theories, exposed frauds, and protected the public health. The best science had to offer to the public could be seen in clear relief at the NDAES.

But there was a negative side to this phenomenon as well. Within the scientific community the sun was setting on the generalist and rising on the specialist. When Henry Bolley left Purdue he could know most of what there was to know about botany. In the next quarter century there was such an expansion of knowledge in that field that a newly trained botanist could reasonably expect to master only a fairly narrow segment of it. Even while Bolley was exercising his remarkable powers of observation to hypothesize the cause of flax wilt it was becoming apparent that future breakthroughs would come from more highly trained and narrowly focused specialists. This is not meant to discount the achievements of the early scientists. They were energetic and idealistic men who did necessary and honorable work that benefited the state and its people immensely. But with the exception of Bolley they did not do work that advanced science or led to the discovery of basic principles. Moreover, the expansion of knowledge in scientific agriculture made it decreasingly likely that they would ever do so. That was the sort of work that required specialists, and the future of agricultural research in this country belonged to them. Like stars that are about to die, the early researchers blazed brightest just before they went out.[42]

There is no indication that Worst and his busy subordinates even recognized these changes, much less worried about them. Their goal was to make the Station a vital, useful, and respected institution in North Dakota, important enough to discourage the John Haggarts of the world from trifling with it. In this they were remarkably successful. But in the process of popularizing the Station they raised up more powerful enemies. Bankers, elevator men, millers, the railroads, seed dealers, food processors, pharmaceutical firms, and others had been bruised pretty regularly by the Station during Worst's directorship. Ultimately they counterattacked, presenting a challenge more serious than any ever offered by the gang politicians of the fading era.

Horace Stockbridge, about 1895.

The College Building, now Old Main, about 1895.

Henry Like Bolley (1) during his 1903 flax trip to Russia.

Edwin Fremont Ladd, 1906.

Experimental milling facilities, 1918.

John Henry Shepperd, in work boots, about 1915.

Longtime herdsman D.J. "Daddy" Geiken with one of the Station rams, 1919.

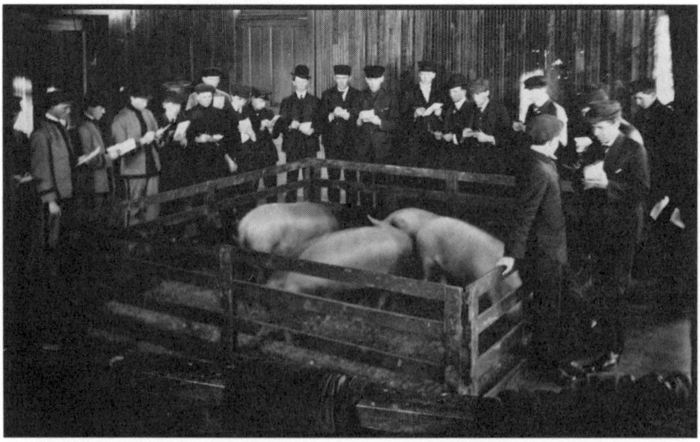

One of Shepperd's livestock judging classes, about 1910.

Picnic of the New Salem Holstein Breeders's Circuit, one of Shepperd's proudest creations, 1916.

Clare Bailey Waldron, 1919.

John Henry Worst, about 1910.

Seeding Station Plots, about 1915.

Cutting corn for Station sileage, about 1910.

Station dairy barn, 1917.

Dairy production room, 1917.

4

THE GREAT CONTROVERSY

No educational institution can be made the subject of controversy and center of this kind of a storm without injuring it.

Fargo Daily Courier-News
August 29, 1914

There is a cancer inside the Agricultural College.

H. L. Bolley to C. W. Buttz,
February 1, 1916

There is a tragic pattern in the affairs of men that is repeated so regularly that it might almost be classified as an essential component of the human condition. At our moment of greatest strength and success, at the instant of our most remarkable triumph, our downfall begins. So it was with the North Dakota Agricultural Experiment Station.

In 1913 the NDAES was a vibrant component of a young and growing state. Its research and regulation touched every man, woman, and child in North Dakota. Its major figures—John Henry Worst, Edwin Ladd, Henry Luke Bolley, and John H. Shepperd—were widely respected and even loved in the state. Its growth in stature had been paralleled by increasing political strength and independence and by dramatic advances in appropriations and numbers of personnel.

But beginning in that year a series of events unfolded that had disastrous consequences for the NDAES. Within a few short years the Station was torn by seemingly endless turmoil. The names of leading scientists were dragged through the mud, and even their research was called into public question. Younger researchers were caught up in a bitter controversy they could scarcely avoid in the hothouse environment of the Station. Some were fired and others resigned with the blessings of their superiors. One director was forced out and two more departed voluntarily in scarcely five years' time. Old friends and colleagues, long engaged in mutual endeavors, attempted to sabotage one another and ceased even the common courtesies of civil conversation. The Station became the plaything of divergent political interests, reaping the distrust of much of the state in the process. In the early stages of the controversy the *Fargo Daily Courier-News* forwarded the opinion that "no educational institution can be made the subject of controversy and center of this kind of a storm without injuring it." That prophecy was accurate, because the NDAES felt the direct effects of the controversy for a generation. And the Station heard echoes of it as long as such principals as Bolley and Shepperd and even such secondary figures as Casper Nelson and O. A. Stevens lived. And veterans of the NDAES live—and remember—for a long time.[1]

What was the controversy all about? For many of the principals and interested observers around the state it was as simple as black and white—only they could not agree which was which. For H. L. Bolley the "cancer inside the Agricultural College" was caused by the attempt of big business, in the guise of the Better Farming Association of North Dakota, to seize control of the institution and prevent it from serving the interests of the people. For others, such as Shepperd, the problem was colleagues who could understand neither what North Dakota agriculture needed nor the direction agricultural science was taking.[2]

However it was perceived by participants, the controversy was inextricably entangled with the intensifying political struggle in North Dakota. On one side were the American Society of Equity, the Nonpartisan League, and others that saw economic cooperation and political reform as means of improving farmers' lives. On the other side were conservative business interests that saw better farming as agriculture's salvation. The NDAES controversy cannot be understood outside the context of the bitter and fundamental political conflict that wracked North Dakota during the teens.

But the divisive state political struggle was not the sole contributor to the great NDAES controversy. The dispute over the Station also involved basic questions about the social role of land-grant institutions and of scientists in a free society. It involved national assessments of what agriculture needed, of what experiment station scientists should do, and of how they should relate to farmers. It even involved questions of optimal organization and direction of scientific institutions and basic differences regarding accepted scientific practices, language, and principles. Personalities played a part in the controversy as well. Self-image and personal ambition became enmeshed with questions of policy. Resentments—by colleagues of recognition given others, by older and younger researchers of one another, between scientists in well-established and newer disciplines—also contributed to the controversy and its acrimony.

The NDAES controversy was a complicated matter. It is difficult to grasp fully even with the benefit of hindsight and was impossible for the principals to understand perfectly. To begin to understand it ourselves, we must start with the role the NDAES chose to play in North Dakota in 1913 and the ways in which that role seemed inappropriate to national leaders in agricultural science.

I

The North Dakota Agricultural Experiment Station—and the North Dakota Agricultural College with which it remained closely identified—chose to play an active and vigorous role in the life of the state during the early years of the century. Part of the decision to do so emerged naturally from the dominant personalities of the early Station. Bolley, Ladd, Shepperd, and C. B. Waldron were men who came to Fargo when the Station and the state were young. They grew up with both, shaped the former, and identified increasingly with the latter. Their public interest and activism were encouraged by College president and Station director John Henry Worst. A successful politician with enduring political ambitions, Worst saw a high public profile for the station as beneficial to it, the state, and himself.

The sort of activism displayed by the NDAC and NDAES under Worst was reflected elsewhere in the national higher education community. Many thoughtful people in this era of progressive political reform believed it was the duty of college faculty to use their knowledge to cure social ills.Many also believed that science could solve all problems, even those that were social or political in nature. The idea of social responsibility and activism was especially pronounced at the land grants because it conformed with their traditional commitment to public service. The University of Wisconsin became famous as the leading activist institution in the country, but others, such as the NDAC, also eagerly embraced a broad social role.

Following their own inclinations, encouraged by Worst, and inspired by contemporary ideals regarding the roles of scientists and institutions, the early leaders at the NDAES involved themselves in a wide variety of activities, some of which were only distantly related to their scientific specialties. Horticulturist-entomologist Waldron, for instance, became a local leader in the socially oriented Country Life Movement. Agriculturist Shepperd ran the Station farm, promoted crops of local importance, and helped create cooperative breeding circuits and marketing structures. H. L. Bolley, a botanist by training, dabbled in virtually every crop, served as state seed commissioner, and became a vocal advocate of economic cooperation in agriculture. Chemist Ladd conducted meteorological observations, paint investigations, and milling and baking studies. He also served as state food commissioner and as a general advocate of increased government regulation in the public interest, and he published a farm newspaper, the *North Dakota Farmer and Sanitary Home*. The early scientists carried these varied interests to the public in their roles as researchers and teachers and also as speakers and popular authors who were in high demand. These were not ivory tower scholars. They were activists involved in the important public affairs of the state.

To some extent the early scientists were so active because North Dakota was then what it remains today: a place with too much to be done and too few people to do it. The state could not afford the luxury of specialists in its early years. Jacks-of-all-trades were more valuable than masters of a few. Scientists had to research, teach, and interact with the general public. The shortage of experts meant that the word of a Station scientist carried a lot of weight, even if he was an entomologist talking about the rural home or a botanist urging economic cooperation. Theirs was a demanding role, but the early figures seemed to relish it. In part they were trained for it, having gone to college when the sciences were ill-defined and when scientists were expected to do everything. But they were also deeply committed to the young and malleable state and were eager to partake in its affairs and help shape its future. They also liked the variety and excitement of their jobs, the public attention they received, and the opportunity to be large fish in a small pond. Shepperd, who lusted after recognition, Bolley, who craved attention and courted controversy, and Ladd, whose apparent shyness and reticence cloaked a healthy ego, all found that their public roles served their emotional needs. The early scientists were what North Dakota needed, but it was what they needed, too.

The scientists at the NDAES enjoyed their activist role, they were appreciated by the state, and they carried out the responsibilities of educated people as defined by many progressive reformers. In the larger agricultural research community, and within its component disciplines, however, very different ideas about the social role of scientists were gaining ascendancy.

As the disciplines of agricultural science matured and became more complex and as basic principles of plant and animal life came nearer scientists' grasp, leading figures in publicly sponsored research began to define a new role for the experiment stations. Increasingly, farsighted scientists conceived that the role of station researchers should be to uncover and develop basic scientific principles, the application of which would result in dramatic increases in agricultural production. This emphasis on basic research was reflected in the passage of the Adams Act in 1906, which provided funds to free scientists from the classrooms and increase their time in laboratories.

The rising emphasis on basic research was paralleled by a tendency to deprecate work that involved direct practical application. "Theoretically it is the primary function of the state experiment stations to conduct researches of a fundamental character which shall be calculated to discover basic natural laws," noted Raymond Pearl, director of the Maine station and an articulate spokesman for the emerging view. But, he complained, "on the contrary, what they do engage in is experimental work of a kind carefully calculated to make as strong an appeal as possible on the basis of its supposed 'practicality' to the scientifically uneducated and uncritical farmers who make up its constituency."[3]

Champions of basic research also argued that regulatory work was inappropriate for the stations. "To require the investigator to lay aside his instruments of research and execute . . . laws . . . is like calling upon an Edison to run a trolley car," contended Director C. E. Thorne of the Ohio station. "The ultimate purpose of the . . . experiment station is to increase the acre's yield, in order that an increasing population may be adequately fed and clothed."[4]

The major problem with regulatory work, from the point of view of Thorne and other like-minded critics, was that it was irrelevant to the pressing social need for more and better food and fiber products from American farms. Basic science was a long-term solution to that problem. In the short term the experiment stations could address it by emphasizing better farm management practices.

Farm management was all the rage in the early years of the century. Farm management specialists believed that farming was mainly a business, and that it could be made more efficient and productive if farmers instituted modern business and marketing practices and took more care in choosing and rotating crops and maintaining soil fertility. In 1902 the USDA organized an Office of Farm Management within the Bureau of Plant Industry. In 1914 that office received separate bureau status. A number of stations organized departments of farm management. Though these usually concentrated on business and marketing issues, some became catchalls that also dealt with soil fertility and rotational research undertaken traditionally by agriculture departments. The emphasis on farm management was intimately related to the rise of county agents, who were perceived mainly as carriers of better farming practices to the countryside.[5]

Those who favored a shift in emphasis toward basic science and farm management did not go unchallenged. Early giants in station research, such as Eugene Davenport of Illinois and Eugene Hilgard of California, who had erected powerful establishments on a foundation of practical research, bitterly deplored the new emphasis on basic work. Cornell's Liberty Hyde Bailey, who saw farming less as a way of making a living than as a way of life, disliked the stress on farm management. But the times were on the side of the proponents of change. Static American farm productivity, a growing population, and consequent dramatic rises in food and fiber costs seemed to indicate that alterations in agricultural research were necessary. Developments in agricultural science seemed to show what was possible, and young researchers, better trained and more sensitive about their professional status than their elders, saw basic research as a means both to making lasting scientific contributions and to achieving higher personal status in the scientific community.[6]

The redirection that station reformers favored demanded a degree of reorganization. Specifically, proponents of the new emphases believed that stations must be separated from colleges and extension agencies and that they must be administered by trained scientists. As early as 1900 the USDA argued that "the station should be organized strictly as a separate department of the institution with which it is connected," and began public encouragement of that separation. The USDA deemed such a division necessary if the stations were to function with full efficiency and a minimum of distractions and political interference. "Successful stations . . . must be managed on the same principles as the great scientific laboratories in the universities are conducted," argued Director A. C. True of the Office of Experiment Stations in 1901; ". . . their officers must be the best trained experts in their respective lines, and they must be able to devote their time and energy quite fully to their investigations." True and others who hoped to alter the direction of research believed station scientists must be largely freed from classroom teaching and from extension work among farmers. The Adams Act was supposed to release researchers from the classroom, and the Smith-Lever Act of 1914, whereby Congress provided support for the county agent system, aimed to relieve them of their extension chores. Scientists would concentrate on basic research, and the structure of their institutions would reflect that fact.[7]

If everything worked out as planned scientists would have more time, but they would concentrate on fewer activities. Proponents of the basic research emphasis urged that both stations in general and scientists in particular specialize much more than was then the case. The stations must concentrate "upon fewer lines and subjects, in spite of the increasing demands for their services," urged True in 1905. "None of our stations are large enough to attack all of the agricultural problems of their localities simultaneously," added E. W. Allen of the OES in 1914. "A selection has to be made, with a view to the needs of science and of agriculture in that locality."[8]

Champions of a redirected and restructured agricultural research system were devoted to the cause of efficiency. The necessity of enhancing efficiency in all areas of public life became an important theme during the progressive period, especially among educated professionals. To their way of thinking, efficiency in agricultural research demanded greater specialization and tighter central direction and coordination. In place of the old, "do-everything" researcher the reformers called for specialized team players who worked under tight supervision in accordance with narrowly drawn plans of work. "No class of men need to readjust their professional code to the modern requirements of the organization of great scientific . . . enterprises more than . . . scientific specialists," True wrote in 1901. "The old regime of the entirely independent . . . investigator has passed away . . . . The fame and emoluments of the individual worker should be subordinated to the requirements of concerted action for a common end."[9]

This conception of research organization came especially from the USDA and conformed to the realities of the largest agricultural research enterprise in the country. It was not a model that was universally admired. Old-style researchers, such as Davenport and Bailey, abhorred it. "Modern efficiency standards are developed from the manufacture of shoes, clothes-pins, overalls, etc., and are expressed in motions per hour," the outspoken Davenport complained. "These standards are not applicable to research." Still, the USDA model reflected the emerging trends of scientific organization, and that agency had a major impact on the practices of the state stations.[10]

No state experiment station could prudently ignore the suggestions of the USDA, which had control over federal funding. Congress appropriated the money, but the OES judged whether Hatch and Adams funds were being expended properly. Moreover, the USDA was in the vanguard of agricultural research, setting examples and delineating standards for the future. Any station that ignored the direction set in Washington would find it difficult to attract and hold the bright young scientists on which its own advancement depended. Top-flight researchers simply would not be drawn to an institution they perceived to be out of step with the times.

Developments at the NDAES in the early years of the century reflected some of the trends of the times. Young specialists with modern training were hired, especially in the rapidly developing fields of plant science. In 1904 the first agronomist was hired, and a bacteriologist and a mycologist (a specialist in plant fungi) came in 1908 and 1909 respectively. There was also a tendency at the NDAES to place a heavier emphasis on the agricultural aspects of farm management, especially by agriculturist John Shepperd. At Shepperd's direction, "crop rotation experiments set up in the early years were reorganized or abandoned," noted Ted Stoa in his history of the Agronomy Department, "and in 1911 and 1912 a new set of soil fertility experiments [was] laid out, patterned after the C. G. Hopkins concepts of soil fertility and crop production." Farm management was the new gospel, and Illinois soil scientist Cyril Hopkins was one of its foremost prophets.[11]

The NDAES, then, was not unresponsive to the currents moving through the American agricultural research community. Its principal figures, however, resisted many of the suggestions coming from leaders in the field. The NDAES showed little interest in basic reorganization. Worst remained the president of the College and the director of the Station, one of a dwindling handful of administrators in the country to hold both jobs. From the modern point of view he was not well qualified for the latter role, since he was not a scientist and he provided little direction to the staff, essentially leaving them on their own to do pretty much as they pleased. Most of the old guard appreciated that system and had thrived under it. "Under your kindly supervision it has been possible for each investigator to follow any ideas arising from any investigations which he may have undertaken to a logical conclusion," noted Bolley in a tribute penned on the day Worst finally relinquished the directorship. But Shepperd and most of the young scientists longed for a professional director.[12]
greater concentration of researchers' energies in the laboratory. Resistance derived largely from the recognition that North Dakota, a poor state, would have difficulty supporting pure researchers who did not teach or undertake extension work. The close identification of College and Station allowed scarce facilities and professionals to be shared. "In my department . . . ," wrote Bolley to Worst when the stricter separation of College and Station was suggested, "it would necessitate the building of extra rooms and perhaps a new building, would duplicate more than one half the apparatus and would in large part duplicate the scientific

workers." Bolley concluded that "in my own department there would be everything to lose to the institution as a whole, and nothing to gain." Much of what Bolley wrote was accurate. Even today, NDAES scientists commonly have substantial teaching responsibilities relative to those at most other state stations.[13]

Worst and the senior scientists were similarly dubious regarding suggestions aimed at emphasizing basic research. Bolley and Ladd clung tenaciously to regulatory work, even though it was increasingly denigrated in the broader scientific community. More than any other factor regulation had helped make the Station and its principal figures popular and important in North Dakota. Surrendering regulation would mean imperiling that popularity and importance. Moreover, it is difficult to imagine anything Ladd could have done that would have served the state better. He was a rather pedestrian scientist, but as a regulator he could make enduring contributions to the state and the nation.

By no means all of the ideas coming down from the scientific elite in the early years of the century were enduring or appropriate for individual stations. The NDAES was wise to be skeptical about them. Nevertheless, it is difficult to avoid the conclusion that much of this skepticism was rooted in a stubborn resistance to the new and different. The pioneer researchers at the NDAES were "do-everything" people. They liked to do everything and they wanted to do everything. When they came out of school they were trained to do everything, and they came to a state where everything needed to be done. But the agricultural research endeavor was changing rapidly. Specialization and organization, not generalization and individual free-lancing, were the wave of the future. Some of the resistance to that future was rational, but much of it was emotional.

Even in the halcyon days of Worst's directorship there were premonitions of trouble. Some scientists believed the Station must move more completely with the trend of the times. Young researchers chafed under the rule of elders they did not respect and became alienated when their senior colleagues shut them out of College governance. Shepperd increasingly bemoaned Worst's lack of scientific competence and derogated Bolley and Ladd as self-advertising dilettantes. Bolley skirmished with the agronomists, continuing to do work that they considered their responsibility. Change was in the air, and as always, it would be marked by conflict and bitterness. The appearance of the Better Farming Association of North Dakota made the conflict dramatically wider and the bitterness infinitely worse.[14]

II

The immediate cause for the creation of the Better Farming Association of North Dakota was an article written by H. L. Bolley that appeared in the *Fargo Forum* on June 26, 1911. In it Bolley indicated that declining North Dakota wheat yields, as well as such conditions as "tip burn" and "sun scald" in wheat, resulted from overcropping in "wheat-sick" soil, that is, soil containing parasitic microorganisms introduced over the years by infected wheat seed.[15]

What appeared on the surface to be the ruminations of a scientific observer called forth a spirited response from people who knew nothing about science. Arthur R. Rogers, a Minneapolis businessman who owned a string of building supply houses in North Dakota, wrote Bolley, "your article has stirred up considerable interest among the business men of this City who are interested in North Dakota and I find there is quite a little agitation among them." Rogers also wrote A. F. Woods of the Minnesota station, who in turn requested more information from Bolley. The latter dismissed those concerned about his theories as "wheat market gamblers" looking to profit from the size of the wheat crop. Representatives of Minneapolis business interests also approached Worst and claimed that Bolley's theory was frightening capital away from North Dakota and harming those with investments there. Director Worst indicated to Bolley that he had heard "considerable complaint from rather influential sources against some of your utterances, relative to the diseased condition of the soil of the state . . . . I am informed that a number of persons are seriously handicapped for the reason that men of means will not loan or invest money in the state for fear that there is something permanently wrong with the soil." Worst concluded by reminding Bolley that "we must handle all matters that effect the financial interests of the state somewhat delicately." Still insouscient about the matter, Bolley replied, "I have no desire to cause even the wheat gamblers to lose money, so . . . I shall try to make it clear in my next public article that a disease infected soil is not necessarily a permanently injured one."[16]

Bolley's theory about wheat-sick soil developed out of his years of observation and conformed to his widely accepted theory about "flax-sick" soil. The major flaw in Bolley's notion of wheat sickness was that it contradicted one of the canons of farm management orthodoxy. Devotees of farm management contended that declining yields were due to deteriorating fertility and poor rotational practices. They argued that better farm practices, such as increased raising of livestock and soil restorative crops, better utilization of animal manures, and more careful rotational practices, would restore fertility and yields.

It was not characteristic of Bolley to bend with the currents of orthodoxy. Indeed, he seemed to revel in confounding the accepted knowledge. He continued to hold to his theory of wheat-sick soil. In a 1913 speech at the University of Wisconsin, reprinted in *Science* magazine, and in *Bulletin 107*, published in December of 1913, he argued that the solution to low wheat yields was pure, sanitized seed, grown in sanitary soil and fertilized with manure free of the parasitic organisms that attacked wheat. Bolley did not argue that better rotation and fertilization were unimportant, but he saw these practices as secondary and insufficient solutions to the problems of low yields. "Wheat deterioration in the fertile wheat growing areas in Northwest America involves soil and seed problems directly analogous to those in the case of the flax crop or potato crop. *It is not primarily a problem of soil fertility lost*, but of disease present," he concluded. Critics came forth to challenge Bolley's theory—or defend farm management orthodoxy—including Cyril Hopkins of the University of Illinois, the great tribune of fertilization and crop rotation. Hopkins wrote that "the persistent efforts to belittle the importance of positive soil enrichment and preservation in permanent rational systems of farming . . . by Professor Bolley, are the greatest curse to American agriculture and the greatest danger to permanent prosperity in this country."[17]

Bolley's theory galvanized men of affairs in the Northwest to action long before Hopkins and other scientists challenged him. In the fall of 1911, a few months after Bolley's theory first appeared in the *Forum*, a bankers' conference was held in Minneapolis on the subject of agricultural development and education. Most of the speakers, including John Shepperd and extension specialist Gordon Randlett of the NDAES, stressed that improved management by farmers was the key to higher yields and greater rural prosperity.[18]

During this conference a private meeting was held regarding the problems of North Dakota agriculture. A. R. Rogers presided and Worst attended. Others in attendance were P. L. Howe, president of the Imperial Elevator company, which owned a string of line elevators in North Dakota, and Fargo bankers E. J. Weiser and Alex Stern. Anger at Bolley apparently had not abated. Rogers reportedly said that "Bolley had done more to injure the credit of the state than any man in it, and . . . that the state could better afford to pay Professor Bolley $25,000 a year to get out of the state or keep his damned mouth shut." Specifically Rogers accused Bolley of scaring Eastern investors away from North Dakota, harming businessmen and farmers alike. Rogers contended that Bolley's actions were one reason why "North Dakota farmers were in bad shape," and he added "that the twin city institution[s] looked upon North Dakota as their customer, and they were willing to go down into their pockets for the necessary funds to render assistance to the North Dakota farmers." Worst suggested that these business benefactors fund an extension system to carry better agricultural practices to the farmer. Thus was born the Better Farming Association.[19]

The Better Farming Association of North Dakota introduced to the state the system of extension education through the use of county agricultural agents. Pioneered by Seaman Knapp in Texas in 1902, the county agent system quickly proved to be a more effective means of conveying agricultural information to farmers than any of the existing devices. By 1911 county agents had become important figures in Southern agriculture and had appeared in a few localities in the North. The USDA enthusiastically supported the concept, and the interest in Congress that eventuated in the Smith-Lever Act in 1914 was developing. The first BFA agent was appointed late in 1911. By 1913 about twenty were in the field, working with farmers to improve agricultural practices.

An undertaking of this scope demanded substantial funding. Supporters initially pledged forty-five thousand dollars. The three main Twin Cities'-based railroads serving North Dakota—the Great Northern, the "Soo," and the Northern Pacific—each contributed five thousand dollars. The balance came mainly from millers, bankers, line elevator companies, lumber dealers, and merchants in Minneapolis and St. Paul. The only individual contributor was railroad magnate James J. Hill. North Dakota sources were expected to match the contributions of the Minnesota benefactors. The North Dakota Bankers Association took the lead in that effort, raising money itself and soliciting local chambers of commerce, commercial clubs, and county commissions. In 1913 the BFA requested legislative appropriations.[20]

The leadership of the BFA consisted mainly of North Dakota bankers, some of whom also had agricultural interests. Eighteen of the twenty-one members of the Board of Directors were North Dakotans, virtually all of them bankers or banker-farmers. E. J. Weiser of the First National Bank of Fargo was president, and Twin Citians A. R. Rogers and P. L. Howe also held prominent positions. Field operations of the BFA were directed by its executive secretary, Thomas P. Cooper, a farm management specialist from the University of Minnesota.

Publicly and privately the BFA always claimed that its primary goal was to help the farmers of North Dakota. As P. L. Howe assured his elevator company's agent in Petersburg, North Dakota, early in 1913, "we are vitally interested in the prosperity of North Dakota." The BFA also affirmed forthrightly that profits for the business interests it represented and prosperity for farmers were complementary rather than contradictory goals. "The interest of the farm is so closely allied to that of the banker and commercial men [*sic*] that whatever benefits the farmer benefits the business man," noted R. S. Adams, president of the North Dakota Bankers Association and member of the BFA Board of Directors. "Mr. Hill has no thought of philanthropy in this particular matter," stated Great Northern president Carl Gray in a remark that must have struck those who knew much about the Empire Builder as an explication of the obvious. "The railroad will benefit by the increased productivity of the lands which it serves."[21]

The BFA proposed to profit both its business sponsors and its farmer clients by making the latter better managers. As Cooper defined them, "the basic agricultural problems confronting the Association may be classified as, the maintenance of the fertility of the soil, the introduction and general use of new economic crops, [and] the business reorganization of the farm for greater profit." When farmers emphasized livestock, instituted rational rotational practices, improved fertility, and modernized their business methods, their operations would become more profitable and productive. And when that happened, more products would flow over the rails and to processors, more money would be spent with merchants and deposited with bankers, mortgages would become safer, and profits would flow in.[22]

This was an attractive and modern vision and a totally attainable one from the point of view of the BFA's sponsors. But many North Dakota farmers were unconvinced. They resented the implication that their difficulties derived from their own shortcomings. Indeed, they believed that the very interests behind the BFA were at the heart of agriculture's problems. The railroads, which overcharged; the elevators, which downgraded grain and docked excessively; the banks, which levied usurous rates; the supply houses and merchants, which reaped unconscionable profits—these were the causes of farmers' problems in the minds of many North Dakotans. This attitude had always been present, and a quarter century before it had fueled the Farmers' Alliance. When the BFA began operating, agrarian radicalism was finding expression in the American Society of Equity, an organization that had operated in North Dakota since 1907.

The Equity's national program called for farmer withholding of crops and aggressive economic cooperation as means of raising prices. Under the strong leadership of state president Theodore Nelson and the charismatic George S. Loftus, the Equity developed a political agenda highlighted by a demand for state-owned terminal marketing facilities in the Twin Cities or Duluth-Superior. The Equity cautiously approved some of the BFA's extension activities, but it expressed a sharply different attitude about the major problem of the state's farmers: "The chief motive of this 'Better Farming' . . . is . . . to divert the farmers' attention from the most important issue, that of a profitable terminal elevator for their farm products," wrote one angry ASE member. "As soon as we can dispose of the one blade at a profit, then . . . we will commence to figure on growing the two blades."[23]

The attitude of the Equity toward the BFA reflected the widening rift between North Dakota farmers and business interests over the nature of the state's agricultural problems and the direction it should take in solving them. Within a few years that division appeared in bold relief with the rise of the Nonpartisan League. In 1911 the nascent farm/business split complicated the response of the Station to the BFA. In a sense, the BFA was nothing more than the most recent and most ambitious of a long string of business—especially railroad—activities undertaken to develop North Dakota agriculture. Promotion of dry farming, stimulation of immigration, establishment of model farms and seed trains, and the distribution of pure-bred livestock were but a few of the steps the railroads had taken to advance the state's agriculture. Most of the time the NDAES had cooperated, usually enthusiastically. The Station was not lavishly funded, and it could not afford to refuse private help. Not only the railroads but also the paint and varnish manufacturers and others partially subsidized the Station. "We ought to be willing to work with anybody who is willing to help agriculture," wrote Bolley in 1912, stating the Station's common response to offers of support from business. Moreover, the BFA seemed to share some of the Station's goals. It wanted to make farmers better managers and to encourage them to raise livestock. Shepperd especially favored those things, but nobody at the NDAES was against them. The BFA offered help in carrying the work of the Station to the farmers in a way that it could not afford. "The Association will cooperate closely with the North Dakota Agricultural College . . . . in taking better methods to farmers," promised an early BFA prospectus. Indeed, Worst himself had suggested this function at the fateful Twin Cities meeting in the fall of 1911.[24]

The BFA received a good deal of cooperation from the NDAES. W. B. Richards, Leunis Van Es, R. C. Doneghue, and other farm management and livestock people were especially helpful. John Shepperd was more than cooperative. Ambitious for the directorship of the Station, then for the presidency of the College, he criticized Worst in "Personal and Confidential" letters to Cooper. Apparently Shepperd believed that his sympathy with the BFA's purposes would put him in good standing with its sponsors. They were men who might have much to say about who ran the NDAES in the future, and Shepperd wanted to run it. Greasing the skids for Worst would not harm his chances.[25]

Shepperd suspected that Worst was cool to the BFA, and he was right. The director strongly distrusted Thomas Cooper and the organization he headed. Worst did not like what he saw as the tendency of the BFA to take credit for work the Station had done. "The work of the Better Farming Association is in the nature of rivalry and direct competition," he complained to the OES in the spring of 1913. He was irritated also that the USDA, which was promoting both farm management and the county agent system of extension, had provided financial aid to the BFA: "The Department of Agriculture has given the Better Farming Association $7500, together with the franking privilege and the assistance of some men, but claims inability to cooperate in any way with our Government Experiment Station for want of funds." Worst concluded by reminding OES chief A. C. True that "the Better Farming Association is a private propaganda financed by the lumber, railroad and elevator interests, chiefly, one one side, and the bankers' associations on the other."[26]

There was more than jealousy here though. Worst and his supporters at the Station may have sensed that the BFA was part of an effort fundamentally to redirect the work of the NDAES or at least to close off some of its traditional avenues of public involvement. One would have to be incredibly obtuse to ignore the connection between criticism of Bolley and the birth of the BFA. Through their regulatory and public advocacy activities Bolley and Ladd had made many friends in North Dakota, but they had also made powerful enemies. Ladd's durum wheat advocacy, milling and baking studies, and pure food and drug work had roused the ire of many in the business community. So too had Bolley's seed inspection work and his public enthusiasm for economic cooperation among farmers. The BFA could reasonably be interpreted as a sign of a more aggressive corporate reaction to what was being done at the Station. Another sign of the times appeared at the end of the legislative session in March of 1913, when Governor L. B. Hanna, who was, perhaps not coincidentally, a major figure in the North Dakota Bankers Association, vetoed appropriations to recompense Ladd for litigation costs incurred as food commissioner and to fund Bolley's work of seed inspection and analysis.

Some of these ominous shadows began to assume a clearer shape in the late summer and early fall of 1913. Events began to move quickly when Worst resigned as director of the Station, effective December 31. He had expressed a desire for some time to concentrate on the College presidency, and he faced at least implicit pressure, internally and from the USDA, to relinquish the Station post. He was also preparing to challenge incumbent Asle Gronna for the Republican nomination to the U.S. Senate, an activity that would necessarily demand much of his attention. The Board of Trustees, dominated by members of the North Dakota Bankers Association, accepted Worst's resignation and then did an extraordinary thing. With the approval of Governor Hanna, the board merged the Station with the Better Farming Association! The former assumed all of the obligations and assets of the latter, and BFA employees became employees of the Station. Cooper was made director of the Station and of extension work in North Dakota, and the BFA's directors were transformed into a board to advise him and the trustees. This merger of a private with a public agency, done secretly by an appointive body, was unprecedented in North Dakota and quite possibly illegal. It was accomplished by men who spoke the same language and who knew what they wanted. They thought they were getting what they wanted when Cooper assumed his new job on January 1, 1914. The great agony of the NDAES was about to begin.

III

Even if Thomas Cooper had not come from a background of corporate influence, and even if he had assumed the directorship at the NDAES under less suspicious circumstances, his tenure would have been controversial. That is so because Cooper represented the new orthodoxy in agricultural science and he was coming to an institution in which many leading figures rejected that approach.

Cooper believed in a fundamental redirection of the NDAES along the lines being drawn by the dominant segment of the agricultural research community. He hoped to bring about a strict separation of "research work from . . . teaching and extension," arguing the "such separation . . . will increase the efficiency of the scientific staff and assure a grade of research work that cannot be otherwise obtained." He favored more clearly and tightly drawn plans of research work and greater coordination of scientists' efforts by administrators. He wanted bulletins written for a scientific audience. He believed researchers should devote their efforts "to a single project, or at most to two projects." And he had clear ideas regarding what the Station should *not* do: regulatory or "control" work was at the top of his list. While he admitted that regulatory work "has offered an opportunity to bring about contacts with the people and to develop departments," he believed that it deflected researchers from the basic scientific work on which they should be concentrating: "With the development of control work resources became strained and laboratory room, equipment and service funds diverted, often to the detriment of research work for which the experiment station was organized."[27]

Director Cooper continued to stress the BFA theme that the farmers' major need was to become better managers. They had to diversify, especially by raising more livestock, attend more carefully to soil fertility, and initiate rational rotational schemes and improved business practices. "Farm profits are ordinarily affected to the greatest extent . . . by the individual practices and methods of the farmer," Cooper argued. "He determines his farm organization, the efficiency of operation and the methods of production which shall be used . . . . The development of each phase of [the] farm operation is dependent upon the individual." When he was pressured to do so Cooper was willing to agree that the Station might appropriately study marketing problems or cooperation, but he always emphasized that the salvation of agriculture was the responsibility of the individual. "Increasing the profits on an individual farm in the state is the problem of each farm owner . . . ," he wrote. "There are . . . two methods by which the income of the farm may be increased:—the first through efficient farm practices and organization; the second by securing increased prices for the products. The first method of increasing income is in the hands of the farmer. The second method is a world problem which the individual affects but slightly." However well this view conformed to that of the mainstream agricultural research community, and however reasonable it seemed to the Minneapolis businessmen and North Dakota bankers who were Cooper's patrons, it was regarded with suspicion in the state. In part because of the investigations of Bolley and Ladd, many North Dakotans believed that the efforts of the individual farmer were undercut by avaricious grain dealers, millers, railroads, and seed companies. Cooper was carrying the banner of farm management into an increasingly hostile country.[28]

Cooper's efforts to reorganize and redirect the NDAES ran up against the determined opposition of Henry Bolley. To the agrarian movement in the state that veteran researcher became the symbol of resistance to corporate control of the Station. To Cooper he symbolized everything that kept the Station from being a modern scientific institution.

Bolley was similar to the better-known Edwin Ladd in the sense that he was a do-everything scientist with a high public profile. Bolley's forte was flax and its diseases, but he dabbled in many other crops and showed little regard for rigidifying disciplinary boundaries. He was an evangelist for pure seed, and his efforts in that area led to the creation of the State Seed Laboratory, which he headed. He was also an advocate of cooperation among farmers and a frequent and outspoken critic of agribusiness interests. Because he failed to see farm management as the panacea for agricultural problems and because his Lamarckianism placed him firmly outside the Mendelian mainstream that held most plant researchers, he was viewed as a scientific heretic. As controversial as Ladd but more vulnerable, Bolley drew most of Director Cooper's attention.

Cooper attacked Bolley's position in a variety of ways. Not only did he separate the State Seed Laboratory from the Station and forbid the expenditure of experimentation or extension funds for seed work, but he took control of Flax Development Committee contributions that Bolley had always handled, preventing use of these monies for pure seed work and eventually refusing them to Bolley entirely. Claiming financial exigency, Cooper also cut Bolley's operating budget, totally eliminating his travel funds and forcing him to dismiss some assistants, and attempted to cut the position of O. A. Stevens, Bolley's assistant in charge of seed inspection. He ordered Bolley to confine his efforts to flax and surrender his wheat work to the agronomists. He gave instructions directly to Bolley's staff, eliciting a protest from Bolley. He raised questions about the quality of Bolley's work with Mark Carleton, chief cerealist at the USDA, and with other federal officials. And he ordered that "all bulletins, circulars and press notices" from Bolley "be submitted to the Director for approval" prior to release.[29]

All of these actions could be justified on the grounds of scientific reorganization and redirection of the NDAES. But Bolley was certain that Cooper was merely the front man for business interests that wanted to drive the botanist out because of his championship of the farmer. Bolley never doubted, as he once wrote his friend and fellow flax researcher M. L. Wilson, "that Thomas Cooper came up here instructed to remove H. L. Bolley, E. F. Ladd, and J. H. Worst before July 1st of the year he first accepted the job," but he lacked conclusive evidence for that belief.[30]

The controversy soon spread beyond Bolley and his work. Worst also became embroiled in a dispute with Cooper. Part of the friction between them derived from the fact that when the Board of Trustees made Cooper director of the Station it did not clearly delineate authority between him and Worst. He became director and Worst remained president of the College, but both were given what appeared to be equal authority and neither was placed over the other. Worst believed he should be the ultimate authority over the institution as a whole, but the board had not given him that mandate. In part because of this ambiguous administrative structure, a bitter and sometimes petty dispute began. Cooper attempted to separate as many quality researchers as possible from College work and confine their efforts to experimentation under his direction. Worst tried to take scientists from Cooper and the director's supporters to augment the College's teaching staff. At one point he created the departments of Agricultural Economics and Agricultural Engineering, removed them and their faculty from the control of John Shepperd, Cooper's most prominent partisan, and attached them to the College. When they could, Cooper and Worst dismissed supporters of the other or made them see resignation as a desirable alternative to working at Fargo. By mid-1914 a distasteful conflict had enveloped the institution.

Worst's resentments, and perhaps his loyalty to Bolley, along with the board's ineptitude, help explain his conflict with Cooper. But it should never be forgotten that Worst was an accomplished politician sensitive to the winds of public opinion. Through no fault of his own he was caught in those winds in

1914. On January 23 Worst hosted the annual meeting of the Tri-State Grain and Stock Growers Association, the support group he had created early in his tenure at the Station. He agreed that, after the regular meeting, the American Society of Equity could have the floor to speak for its idea of a state-owned terminal elevator complex. Almost immediately, members of the Fargo and Minneapolis commercial clubs in the audience began heckling the Equity speakers, and after a time these businessmen and members of the Fargo Police Department broke up the farmers' meeting.[31]

Worst found himself in a very difficult situation. Equity members accused him of duplicity for promising them the floor and then failing to stand behind them. The Fargo businessmen detested everything that Equity stood for and they did not appreciate Worst's giving the organization a forum. Worst's position was especially sensitive because he was running for the Senate. Attacks by the Equity weakened him and contributed to his loss in the primary to Asle Gronna. Like most successful North Dakota politicians, Worst had always tried to maintain good relations with farmers and businessmen both, but that was becoming more difficult to do in an increasingly polarized political environment. "Pres. Worst had one foot in the Chamber of Commerce and one foot in the equity" noted one observer at the NDAES. As the gap between them grew he had to jump, and he jumped to the side of organized agriculture. That decision further intensified the conflict at the NDAES.[32]

By the late summer of 1914 the controversy involving the College and Station had entered the public arena, where attention focused primarily on two issues—the legality of the merger between the BFA and the NDAES and the purposes of the former. Critics of the BFA argued that the merger was illegal, mainly because it obligated the state to expenditures that were not yet authorized by the legislature. Supporters of the merger argued, correctly, that because of carryover funds in the BFA account the state was not immediately obliged to spend public money. The merger did, however, implicitly commit the state to future expenditures. Defending the merger, the *Fargo Forum* argued that it was necessitated by the Smith-Lever Act, which provided federal funds for state extension programs. The Smith-Lever Act was not passed until several months after the merger was arranged, but the *Forum* was determined not to let the facts get in the way of a good argument. Governor Hanna offered to investigate the legality of the merger, but the Equity paper saw the hypocrisy of that suggestion. "The governor himself is the biggest toad in the puddle," snorted editor George Nelson of the *Co-Operator's Herald*.[33]

The major point at issue, however, was the BFA's purposes and its goals for the Station. The BFA always claimed that its objective was simply to help farmers become better and more productive. In order to do that it attempted to advance among them improved methods of production and business. This aim, with its implicit suggestion that most of agriculture's problems were located in the farms themselves, was controversial enough in the context of North Dakota in 1914.

But critics professed to see in the BFA a more pernicious, secret agenda. They saw it as an attempt by big business to control higher education in North Dakota and to stifle those who exposed corporate maledictions. To the mind of Equity editor Nelson, what was afoot was "one of the most insidious, subtle and far reaching conspiracies against the common welfare that has ever been attempted in this or any period of civilization." The BFA had as its goal changing the College and Station from an institution performing "a service along agricultural lines similar to that of the University of Wisconsin along social and industrial lines of reform" to one subservient to corporate interests. The first step in this effort was the removal of Worst, who had "been too liberal with the faculty and student body and [had] given them too much latitude in the expression of views and convictions that do not square with the orthodox views of the powers that be." The next step was the removal of Bolley, Ladd, and others who had roused the ire of big business.[34]

Nelson's thesis about the BFA's agenda seems conspiratorial and unsophisticated to the modern reader, but it is not totally implausible. After more than a decade of being blamed by progressive reformers for everything from poverty and the high cost of living to the deterioration of American moral values, and after a period of increasingly stringent government regulation of most spheres of business activity, corporate America was striking back in the early teens. Its weapons were economic, and it was not reluctant to use them. Advertisers withdrew business from muckraking magazines. Wealthy parishioners abandoned churches led by crusading ministers. Philanthropic support for social reform organizations dried up. And activist college professors came under increasing pressure from boards composed of wealthy trustees and from conservative legislatures. The fact that this was occurring elsewhere does not prove it was happening at the NDAES. But it is important to remember that the people behind the BFA were used to having their way in North Dakota and that they were not known for their commitment to the abstract ideal of academic freedom.[35]

But what Cooper and the BFA were really up to—what their hidden agenda entailed—remains an intriguing question and one to which existing sources do not permit a conclusive answer. It is clear that the people bankrolling the BFA disliked the activities of Bolley and probably Ladd as well. It is clear that Cooper remained respectful of the wishes of his wealthy supporters even after he became director. It is clear that he made life miserable for Bolley—and for Ladd, albeit to a lesser extent—and it is probable he would have viewed Bolley's exit with favor. It is also clear that Cooper was determined to rule the Station, even if that meant removing men who shared his ideas about the problems with the state's agriculture, and that his corporate patrons were willing to lend a hand in that effort. Extension specialist Randlett, for example, was given no responsibilities by Cooper, even though he shared Cooper's publicly stated goals. Director T. A. Hoverstad of the North Dakota Farmers' Institute learned years later that the BFA engineered his removal, despite his orthodoxy on farm management. Hoverstad wrote to his friend John Shepperd that he had been told by a figure prominent in the BFA that "you were offered the Agricultural Dept. of the Soo Line . . . simply to get you away from North Dakota. This would give the B.F. Assn. full controll [*sic*] of the Farmers' Institute." Despite the fact that there is no smoking gun in the sources, there are enough whiffs of powder to indicate that the wariness at the Station and in the state regarding the BFA was fully justified.[36]

Even if they did not shape specific events at such institutions as the NDAES, it is important to remember that private interest groups exerted influence over the general direction of research. The blatant control of scientific endeavors and theoretical constructs in such societies as Stalin's Soviet Union and Nazi Germany has been well documented. But it would be unrealistic to believe that science in freer societies such as the United States is totally uninfluenced by powerful social groups. Sometimes the majority determines the direction of scientific endeavor or forecloses certain potential avenues of research. Today, for example, our social commitment to equality effectively dissuades reputable scientists from investigating possible genetic differences related to race or ethnicity. It is also true that even in relatively pluralistic societies some groups have more influence than others. In capitalistic societies corporate interests traditionally wield disproportionate power. They influence science because many scientists work in the private sector and because they help determine the flow of resources into the public sector. When they support legislative appropriations or make direct grants, corporate interests can be expected to favor some areas of scientific endeavor—production-oriented agricultural research, for example—over others, such as regulation. This is usually a subtle process—imperfectly understood by both the influencer and the influenced—and there are often countervailing forces involved, but it is a fact of scientific life in a democratic capitalistic society.[37]

The exercise of corporate influence over agricultural science in general was too subtle for most people at the NDAES to grasp. But even if they had fully comprehended its dynamics their behavior would probably have been the same. Cooper justifiably believed he was taking steps to make the Station a conventional agricultural research institution according to the standards of the time. Tighter direction and control over scientific endeavor, separation of research from teaching and extension, avoidance of regulatory responsibilities, a laboratory emphasis on basic work and a field emphasis on farm management—all of these conformed to accepted practices in the national agricultural research enterprise in general. We will never know whether Cooper was motivated primarily by a desire to bring the Station up to date or by blatant big business toadyism or by subtle corporate pressure, but the source of his actions is not terribly relevant. He, or any other young scientist aware of the direction of research, would have pursued the same policies in any event. What suspicions involving Cooper's purposes did was to make those policies and their implementation much more painful and controversial for the Station and for the state.

The controversy over the Station ended up in the lap of the Board of Trustees, which tried to grapple with it in the fall of 1914. It had to do something. The issue had become public and could no longer be ignored. The Board perceived Worst to be the biggest fly in the ointment. He was undermining Cooper, sometimes in petty ways, and board members suspected him of leaking embarrassing information to the *Co-Operator's Herald* and other critical papers. The problem was that the board could not simply fire him without giving credence to the charge that it was the tool of big business. In November, while Worst was out of town, the Board fixed on the ingenious solution of making him "president emeritus" and making Ladd president of the College. Ladd agreed, on the conditions that Worst was amenable and that the board would give Ladd primary authority over the whole institution; Worst wasn't, and the Board wouldn't, so the scheme fell through and the controversy continued to fester.[38]

The members of the board were probably relieved when, early in 1915, the legislature decided to change the structure of governance in higher education. A new Board of Regents was created with new members appointed by the governor. But changes were more apparent than real. Like the old board, the new one was dominated by members of the North Dakota Bankers Association, which Governor Hanna seemed to perceive as a special repository for wisdom and competence in the state. L. F. Crawford, a Sentinel Butte banker and past president of the NDBA, chaired the new board. Hanna was obviously comfortable with a banker-dominated board, but its complexion made the public suspicious of any actions it took regarding the principals in the burgeoning Station controversy. As subsequent events showed, the composition of the board severely limited its freedom of action and the number of options open to it.

Something of a climax in the debilitating struggle at the Station came in the early weeks of 1916 through the agency of Bolley and Worst. On January 21 Bolley wrote board chairman Crawford, asking that the board give him permission to reorganize his department. Apparently he intended to fire subordinates who sided with Cooper. As he explained later to Worst, he wanted "to dismiss any employees . . . necessary to bring about harmony in the work of the Department." A week after writing Crawford, Bolley publicly aired the Station's dirty linen at an agricultural fair in Devils Lake. Bolley used a parable to thinly veil his contention that Cooper, the corporate timeserver, had caused all of the problems at Fargo: "Once there was a great prosperous, co-operative farm, successful in every way, and a model to all who came to see it," he told the stunned audience. "There came a time when the managers sent to the yard a boy, clothed in an invulnerable jacket of authority. He worried the animals, kept the bees stirred up. Some of the animals were stung. Some of the cattle and horses broke their harness. Others ran into the wire fence and killed themselves. Some went through the gates and ran clear out of the country."[39]

Then, on February 18, Worst "requested," or "demanded," that the board extend his contract for a year and give him authority over both the College and the Station. He said he needed this grant of power to "whack a few heads" and "torpedo some submarines" on campus. Worst and the board had reached the end of the line. Taking a cue from their predecessors, the regents appointed Worst "president emeritus" on February 19. Now the firestorm struck. Students and professors rallied to the beloved "Prexy" and lambasted the board in public meetings. The board came to Fargo to put students and faculty back in their places, only to be criticized to their faces—along with Cooper—as minions of big business. They had to get a new president, and quickly, but who was it to be? They seemed to prefer an outsider, but what quality educator would be willing to step into the situation at the College? The *Fargo Daily Courier-News* had reported the rumor that Treadwell Twichell was to be the new president. But appointment of the legislator who supposedly advised farmers seeking a state-owned terminal elevator to "go home and slop the hogs" as president would be an act of political stupidity monumental even for a governing board in North Dakota higher education. No, this bankers' board had only one option in the prevailing political climate. It turned to Edwin Ladd. Realizing he was the necessary man, Ladd demanded control over the whole institution. The Board had no choice but to give it to him.[40]

Everybody praised the appointment of Ladd, with the conspicuous exception of Cooper, who offered no comment. The pure-food crusader was probably the best-loved figure in the state in 1916. The *Forum* editorialized that "the people of North Dakota have the utmost confidence in Dr. Ladd," and the *Co-Operator's Herald* added that "almost without exception the people of North Dakota admire and respect" him. Even the *North Dakota Banker* lauded the choice, adding hopefully that "the appointment . . . puts at rest the silly talk of the Board of Regents being influenced in their actions by anybody or any motive other than that of doing what is best for the College and the taxpayers of the state." Bolley wrote a Purdue dean that those desiring "to control the educational institutions for their own private purposes will run into a real snag when they undertake to do so thru Prof. Ladd." One of Shepperd's correspondents noted that "Prexy got his'n finally—Why Waldron and Bolley have not got theirs long ago I cannot understand. I hope Ladd will live to straiten [*sic*] out the tangle." Some of the faith in Ladd derived from genuine gratitude for all he had done for North Dakota and from relief that the debilitating controversy might be ending. But that so many people with diverse viewpoints could believe in him also indicated that he kept his own counsel. People tended to attribute their own hopes and ideas to him, assuming he shared them. Sometimes he did and sometimes he did not, but he usually would not reveal himself either way. When he assumed the presidency he was more a man of mystery than many people believed, and he continued to be so for some time.[41]

Those who hoped Ladd's appointment would end the controversy were soon disappointed. A movement was afoot to remove Bolley from his position on grounds of scientific incompetence. The vehicles for this proposed removal were two disgruntled researchers in Bolley's department, W. H. Mercer and D. G. Milbrath, but Shepperd stood behind them, and Cooper almost certainly stood behind him. The charges against Bolley went back to the precipitating cause of the whole BFA affair—his theory that declining wheat yields were at least partially attributable to the presence of parasitic microorganisms in the soil. Mercer and Milbrath leaned heavily on Hopkins in charging that Bolley's research was sloppy and that he drew erroneous and unwarrantable conclusions from the evidence he presented. Even more serious was their accusation that he used misleading photographic and statistical evidence and fabricated results in *Bulletin No. 107*.[42]

One need not be a champion of Mercer and Milbrath to see an element of truth in some of their charges. Bolley always had so many irons in the fire that his work was sometimes slipshod. He admitted to making errors in the preparation of *Bulletin No. 107*, including failing to have many notes on record. He argued that the bulletin was hypothetical and that, as was the case with all of his work, it was written for his employers—the farmers of North Dakota—rather than for scientific researchers. Whether Bolley faked the work is another of those unanswerable questions that abound regarding this controversy. It is clear that he was under great pressure when he was preparing *Bulletin No. 107*. It was put-up-or-shut-up time for him, and it is not inconceivable that he fudged to support his points.[43]

Bolley believed that the charges had more to do with politics than with science. He thought Cooper and Crawford and the interests behind them were out to get him because he was politically obnoxious, not because of his scientific deficiencies. In the early months of 1916 he gathered what evidence he could about his accusers, writing friends in the USDA and in other stations and pressuring his subordinates. The struggle assumed a bitterness comparable to that in the war then going on in Europe. And just as in the trenches in France, in the offices and labs in Fargo the privates were more likely than the generals to be casualties. Bolley put tremendous pressure on his staff for any information they might have, especially evidence he might use against Mercer and Milbrath. While Bolley was always sensitive about his own freedom, he dominated subordinates with an iron hand. Now he placed them in a painful dilemma. They could betray their colleagues in order to work in an environment of incredible tension or they could deny their superior and face likely dismissal. In the letters of such people as Nevada Evans and Casper Nelson the torment and anguish of the situation was clear.[44]

Ladd's position in this affair was enigmatic initially. But in July he appointed a special committee to investigate the charges against Bolley. Ladd had been at the College long enough to know that one never appoints an impartial committee unless he knows what it is going to decide. To this committee he named C. B. Waldron, E. S. Keene of Engineering, and veterinarian Leunis Van Es. Van Es's factional preferences are not clear, but Waldron and Keene were both Bolley supporters. After due deliberation the committee came up with the expected conclusion. Bolley was exonerated and he quickly dismissed his accusers. The combative botanist was not completely out of the woods yet—the board and the USDA still wanted to look into *Bulletin No. 107*—but he could see open country ahead.

The action of Ladd's special committee helped put the handwriting on the wall for Cooper, and the Nonpartisan League helped him read it. The Nonpartisan League was the cowbird of agrarian politics in North Dakota, pushing the Equity Society out of the nest it had so carefully prepared. Like a "political prairie fire," the NPL swept out of western North Dakota in the early months of 1916. It called for state provision of services and ownership of facilities that farmers used. Ultimately it gave North Dakota a state-owned bank and a state-owned mill and elevator complex, unique institutions in the United States. In June of 1916 League candidates virtually swept the Republican primary, winning designation for every office but one. In November they won an electoral landslide, placing Lynn Frazier in the governor's chair and seizing a majority in the state house of representatives. Cooper was willing to trim his sails before the shifting wind, suggesting that marketing questions and cooperation in agriculture might be fit subjects for investigation by the NDAES. But nobody in the NPL trusted him, believing, probably correctly, that he urged the county agents to work against them in local areas.[45]

In December of 1917 Cooper announced that he was resigning as director to take a similar post at the Kentucky station. Conservatives were crestfallen. With the unconscious irony in which it excelled, the *Forum* warned readers that "men of the calibre of Mr. Cooper cannot be expected to stultify themselves by remaining in state institutions that are the football of a political propaganda, and where they are expected to make their thoughts and utterances conform to the political desires of the men in St. Paul who control their appointments.* His enemies were elated. "Most of our troubles have emigrated to [a] warmer climate," Bolley chortled.[46]

Now Cooper's enemies took vengeance on his friends. Ladd persuaded the Board of Regents to abolish the post of vice-director of the Station, which was held by Shepperd. To add insult to injury he failed even to tell Shepperd what had been done. It was an act of remarkable pettiness, but one that served as an appropriate conclusion to an episode in which there were no heroes.

IV

During the period it was torn by controversy the NDAES was not a very productive research entity. Between 1914 and 1919 only twenty-four bulletins

**The "men in St. Paul" to whom the Forum referred presumably were the leaders of the NPL, which had its national headquarters there.*

were published by the Station, in contrast to the sixty published in the six years that followed. In 1914, Cooper's first year, and 1918, when Leunis Van Es and Perry Trowbridge, in turn, took the reins, a total of only four bulletins were published. Most of the bulletins that were produced either dealt with ongoing projects such as Ladd's wheat studies and reports from substations or contained farm management pieces of a derivative and pedestrian nature. The printed record does not indicate that much work of significance was coming to fruition.

It is tempting to attribute this apparent decline in productivity solely to the controversy at the NDAES and the consequent disruptions there. The atmosphere created was obviously not conducive to serious research, and Bolley, for one, complained bitterly of how his own work had been harmed. The factional struggles that broke out were not only distracting. They also resulted in dismissals and a large number of voluntary resignations. Of the thirty-nine staff members on the rolls in January of 1914 only five were still listed in 1919. In essence, only the major figures at the Station stayed during this trying period.

At the same time, it should be remembered that this was a very difficult time for agricultural experiment stations everywhere. The Smith-Lever Act, which provided a degree of federal support for extension, was supposed to be a boon for the stations, freeing scientists to concentrate on research. In fact state legislatures tended to support county agents out of funds that would otherwise have gone to the stations. As OES chief E. W. Allen noted in 1920, the result was "a remarkable anomaly—that of providing a vast and growing system for the popular dissemination of information while halting the growth of the agency on which it must depend for its very stock in trade." Total state and federal appropriations for agricultural experimentation had doubled every five years up until 1914. Between then and 1919 they rose hardly at all. The situation was rendered especially serious by the dramatic inflation of the period. Between 1914 and 1919 average wholesale and retail prices nearly doubled. In real terms the stations had about half as much money in 1919 as they had in 1914. The consequences of attempting to stretch static resources to meet increasing costs could be seen most clearly in the area of employment. New opportunities in extension and private business enticed quality researchers away. High turnover disrupted the work, and replacements were seldom as well-trained or experienced as those they replaced. It was a sellers' market in which anybody who wanted to move and was at all competent could go elsewhere. In 1920 the USDA estimated an average station turnover of 80 percent in the previous six years. It was a nice situation for agricultural scientists who craved mobility, but not very good for agricultural research.[47]

Part of the cause for the remarkable disruption of agricultural research was the First World War. The unwillingness of the government to control prices and its decision to partially finance the war by expanding the money supply contributed to inflation. Many patriotic young scientists enlisted in the Army, even though they were commonly granted deferments on request. Others left to work for such agencies as the USDA and the United States Food Administration. Those who stayed at the stations were burdened with demands unrelated to their long-term research interests. Researchers often found emergency war work exciting, but crop reports and food surveys took time and energy away from agricultural experimentation. P. F. Trowbridge's first report as director of the NDAES repeated a refrain heard in most other states: "No increase in funds was available and the continually mounting cost of both service and supplies forced a readjustment and curtailment of a number of projects. The call of . . . men into war service left projects without leaders. Workers who remained at the . . . Station were . . . called upon to divert their energies into lines of activity closely associated with the war emergency." There is no reason to believe the NDAES would not have been as disrupted as other stations were during this troubling period. The controversy simply exacerbated an already difficult situation.[48]

So why was the controversy so important? It cannot be said to have been the primary cause of the disruption of work at the Station, nor did it alter the direction of scientific development at Fargo. That was governed more by national currents than by local circumstances. It was important in part because it divided people who had worked in a mutual enterprise for many years. Decent people behaved in a petty and vindictive manner toward friends and colleagues; they showed their worst sides to one another—perhaps some even came to see the flaws in themselves.

Internal differences can be overcome, and they were at the Station to some degree. A great intangible is the effect of the controversy on recruitment. The Station always found it hard to attract and hold good people because of its locational and financial disadvantages. Now it bore the additional stigma of being a highly divided and politicized institution. At about this time increasing numbers of North Dakota natives began to join the staff. To some extent this reflected the maturity of an institution training a generation of local researchers. It might also have been an indication that outsiders were increasingly difficult to attract.

The major result of the controversy was to politicize the NDAES in a way it had never been before. The Station has always been a political institution, as every successful director has realized. It receives public funds through a process involving legislative and executive bodies, and it responds to the demands of political interest groups. Scientists do not always understand this and they usually abhor it. As Station veteran Ted Stoa put it, "most of us didn't want a darned thing to do with politics; they were so unfair." Stoa's attitude was understandable and widely shared, but it ignored both the realities of publicly sponsored research and the political implications of scientists' work.[49]

The problem with the controversy was not that it involved the Station in politics per se, but that it made the Station and its direction the center of a bitter political struggle that divided the state for a generation. Because Ladd presided, Cooper left, and Bolley stayed, conservatives concluded that the NDAES and the

College were dominated by radicals—"a bunch of socialists," as one of Shepperd's correspondents put it. This was not a very accurate characterization, even of the principal figures in the controversy. Ladd was ambitious and shrewd, but he was not a very sophisticated political ideologue. His nomination by the NPL for the Senate in 1920 and his subsequent election were mainly tributes to his great popularity in the state, a commodity that the League hoped he would transmit to its own faltering ticket. But he was basically an old-style progressive. He conceptualized politics as a simple struggle between most of the people, who were good, and selfish, predatory economic interests, which were bad. He did not agree with all portions of the NPL program, and the League probably would have dumped him from the ticket had not death removed him as a political figure in 1925. Bolley was committed to the NPL and to other progressive causes, and he was often outspoken about his beliefs. He supported the League even during the thirties when it was dominated by William Langer, the arch-tormentor of the College and Station. On another litmus-test issue of the times, he was a vocal supporter of the Soviet experiment even after Russian friends attempted to acquaint him with the horrors of agricultural collectivization and Stalin's purges. But Bolley's radicalism was of the emotional variety. In 1928 he voted for Herbert Hoover for president because Al Smith wanted to modify Prohibition. That was hardly the stand of a thoughtful radical.[50]

Distrust by conservatives, however ill-founded, did not translate into support by radicals for the Station. As long as it was associated with Extension, and Extension was associated with Cooper, the left would distrust it. Moreover, as the Station continued to jettison regulatory work and eschew advocacy on public issues it invigorated radical suspicions of its motives. The Station was influenced by national scientific trends. Its ability to attract quality scientists and hold USDA support with a program stressing regulatory work and public advocacy was distinctly limited. But while it got in step with the national drummer the North Dakota left heard the cadence set by Tom Cooper and the BFA.

The prophecy uttered by the *Fargo Daily Courier-News* in 1914 regarding the effect of the controversy was fulfilled in the twenties and thirties. For a generation North Dakota radicals and conservatives tore at one another like pit bulls, and all too often the College and Station were caught in the middle.

5

COOPERISM WITHOUT COOPER

These Land Grant Colleges put the spirit of righteous fighting into the common people. The time is not here to give over the fight.

H.L. Bolley, March 10, 1921

The College at Fargo . . . so far as I knew, was the only Agricultural College which told the entire truth. It may well be it has under pressures become entirely safe and sane.

Joel H. Greene to H.L. Bolley, Undated

Thomas Cooper's imprint remained even after he left the North Dakota Agricultural Experiment Station. He had been unable to reshape every aspect of the institution, but most of his key ideas continued to inform Station practice during the twenties. The stress on farm management, the opposition to advocacy on public issues, and the avoidance of regulatory responsibilities were some of Cooper's emphases that survived him at Fargo.

The continuation of Cooperism after Cooper was, for the most part, not testimony to the power of his personality. Cooper had carried the message of developing trends in publicly sponsored agricultural science to the NDAES. Those trends, and the responsiveness of the research community to them, did not change just because a particular messenger went away.

Some continued to abhor the direction taken by the research establishment. H. L. Bolley sounded the call for rededication to the principles of public service and activism he believed the land-grant system stood for. "These Land Grant Colleges put the spirit of righteous fighting into the common people," he argued in 1921. "The time is not here to give over the fight." But it was. The public mood had become dramatically more conservative. Research institutions had narrowed their focuses and become more prudent and cautious. Scientists were more professional, more careerist, and more reluctant to speak to issues outside their narrow areas of expertise. To passionate, do-everything idealists such as Bolley, the stations were becoming foreign. Like an alien in a strange land, Bolley spoke to colleagues who did not really understand him. The NDAES became less idiosyncratic and more conventional. In the words of one of Bolley's radical correspondents, it was becoming "entirely safe and sane."[1]

Ironically, the national agricultural research establishment to which the NDAES was trying to conform was suffering its own crisis in the twenties. In the early years of the decade the problem was mainly financial. The stations simply could not rebound quickly from the economic ravages of World War I. "The stations are marking time, on half rations, little new work is being inaugurated, station men are deserting and going into business enterprises, and few trained men are being recruited" lamented Director E.A. Barnett of the Nebraska station in 1919. "More money is needed."[2]

More money was not forthcoming. In 1921 a conservative Congress, frightened by wartime deficits and anxious to balance the budget, cut research funds dramatically. With the country in the throes of a severe business recession, few states were able or willing to increase their own commitments to agricultural research.

Money was not the whole problem. There was also the troubling question of what agricultural researchers should be doing. Almost overnight overproduction replaced underproduction as the main difficulty of agriculture. Rising American production, the rapid recovery of European agriculture, and the reluctance of many of our trading partners to deplete their thin dollar reserves by buying our commodities conjoined in 1920 and 1921 to devastate farm prices. There was some recovery during the twenties, but throughout the decade American farmers produced more than could be consumed by paying customers. Some critics blamed agricultural scientists for the situation and argued that overproduction made researchers superfluous. Others contended the agricultural research establishment should place increased emphasis on such areas as marketing, cooperation, and the rural home and social life. Ultimately, the system devoted more resources to those fields, but its main thrust remained toward enhancing and increasing production.[3]

The twenties was an especially difficult time in North Dakota. While the nation's agriculture had generally thrived during the war, adverse circumstances prevented North Dakota farmers from benefiting fully from high prices. Of the six wheat harvests from 1914 through 1919, only one could be classified as good, and drought and rust rendered three poor. By 1921 wheat had plummeted to barely a dollar a bushel from the $2.25 it had commanded two years earlier. Never again during the twenties did prices approach 1919 levels. Compounding the misery was the fact that wheat yields in North Dakota during the twenties continued to be only fair or poor.

The result of stagnation in wheat was a painful "time of readjustment," in the phrase of North Dakota historian Elwyn B. Robinson. Farm mortgage debt and tenancy increased dramatically. Dozens of undercapitalized and overextended rural banks failed, pulling mortgagees down with them. The state had 898 banks in 1920. Only 325 of those remained in business in 1933. Recoiling from an apparently bleak future, young people fled the countryside in droves, initiating an exodus that continues even today. Farmers who stayed struggled to survive. They lived by strict economy and attempted to mechanize and achieve economies of scale. Hardship on the farm was felt by everyone in the state. As NDAES researchers R. Wayne Newton and Alva Benton pointed out, "the prosperity of the whole State is at present bound up with the prosperity of the farm." As a consequence, "all incomes . . . vary in the same direction at the same time." Whether they lived in town or on the farm, for North Dakotans in the postwar period the boom of just a few years before was but a memory, the promise of which had evaporated like the morning dew.[4]

I

The imposing challenge of keeping the NDAES in step with the agricultural research establishment, stretching scarce dollars to cover abundant needs, and meeting the demands of depressed North Dakota agriculture was assumed largely by a new director, Perry Fox Trowbridge.

When Thomas Cooper departed for Kentucky at the end of 1917 Leunis Van Es was made temporary director; he left to take a job in veterinary science at Nebraska in April of 1918. John Shepperd campaigned vigorously for the directorship, but it eluded his grasp once again. That Shepperd thought he had a chance given the local political climate in 1918 indicated an impressive capacity for self-delusion. The Board of Regents wisely decided to hire from outside, bringing Trowbridge in from Missouri.

Fifty-two years of age in 1918, Trowbridge had degrees from the universities of Michigan and Illinois and was at the University of Missouri when he was tapped for the NDAES job. His academic specialty was animal nutrition. Because he left no papers, Trowbridge's personality and character are something of a mystery. Those who knew Trowbridge spoke of him in generally favorable terms. Chris Jensen's most vivid memory of Trowbridge came from the day the former arrived at the Station: "I went to meet him, and he was in his shirtsleeves, cutting up a calf. I figured he must be a regular fellow. He put me to work helping him." Trowbridge was noted for his hard work, sound judgment, and loyalty. Obviously, though, the man had qualities that are not revealed by vague memories and banal praise. He held the directorship during a time of sustained difficulty. When he came the NDAES was still ripped by internal controversy and burdened with external suspicion. During his early years financial stringency racked both the Station and the state. When he left his post in 1934 the Station was in the midst 16 yrs.

of its most serious fiscal crisis ever. Trowbridge could not solve all of these problems, but he did address them, sometimes in creative ways.[5]

Trowbridge's tenure was marked by general financial exigency. Nationally the research enterprise was reeling from the spare appropriations and high inflation of World War I. Some scientists were hopeful that the difficult situation would be alleviated quickly. Director R. W. Thatcher of the Minnesota station, for example, argued in 1920 that the war had resulted in "a tremendous stimulation of interest in organized scientific research." Unfortunately that supposed interest was not reflected in rising appropriations. The double blow of the business recession and the agricultural depression lowered funding levels. Exacerbating the situation was intense competition for rare dollars from colleges burdened by rising enrollments and extension agencies backed by the new and powerful American Farm Bureau Federation.[6]

The situation in North Dakota was worse than in most other states. The war years had not been remunerative and the agricultural down turn was especially severe in the Flickertail State, dependent as it was on a relatively undiversified farm economy. A dramatic increase in enrollment at the Agricultural College—from an average of around 250 in the late teens to over 800 in 1923—strained finances, and the Station continued to bear the stigma left by the great controversy.[7]

Trowbridge's first years were especially difficult from a fiscal standpoint. Reporting on the NDAES during 1919-20, he argued that "the year has been one of the most serious financially in the life of the Experiment Station. With no increased appropriation during the last ten years and with rapidly increased cost of operation it has been very difficult to maintain anything like normal efficiency." The situation was more serious at the branch stations. Superintendents simply could not hire labor with the available funds. At Dickinson twelve-year-old Glenn Smith, the son of Assistant Superintendent Ralph Smith, mowed weeds in 1919 because nobody could be hired with the money available. This gave him an early suspicion that working for the state of North Dakota was not to be the way to great fortune. In 1921 the Station received a blow from Washington when the USDA withdrew funds that had helped maintain three of the substations. A special state appropriation made it possible to continue work at Edgeley, Hettinger, and Williston at a minimal level. Two years later cuts forced the abandonment of the Demonstration Farms program, of which Worst had been so fond, and ended soil survey work. In 1925 the legislature deeded the Williston substation to Williams County for use as a poor farm. In light of the financial situation of the NDAES at the time, that seemed an especially appropriate use.[8]

Financial exigency forced a retreat at the Station from the grandiose vision of Thomas Cooper. Cooper foresaw a large and specialized research establishment strictly separated from the College and Extension. This plan conformed to that advanced by the USDA, and Cooper did his best to put it into operation. Whether it was ever realizable in a state with as few resources as North Dakota commanded is doubtful. It was clearly not achievable in North Dakota in 1919.

Trowbridge began to modify the Cooper structure soon after he took command. He had little choice. The College and Station resumed sharing of facilities, equipment, and staff. Station personnel returned to the classrooms, and faculty in the College resumed Station investigations on a part-time basis. Trowbridge also worked to rebuild the bridges between the components of the institution that had been undermined during the struggle of the mid-teens.

The USDA was not pleased with the retreat from the Cooper system, because it thought that research work would suffer and that funds would be inappropriately expended. But the NDAES was not the only station forced to respond to financial exigency in this manner, and in any event, the USDA could or would do little to prevent the change. "I think this is a matter which the U.S. Department cannot wholly control, but . . . is properly one of state policy," contended H. L. Bolley. "A large institution can perhaps justify having all of its men on the Station full time and all of its college teachers full time, but I submit that we would not have funds enough to go around on such a basis." Perhaps the USDA could have made life more difficult at Fargo by a narrow and strict definition of appropriate expenditures of funds, but as usual it chose to advise and encourage rather than attempt to dominate.[9]

Reintegration of the Station with the College did not solve the financial problems of the former. Trowbridge continued to struggle to stretch resources to cover basic demands. Wages paid to technical staff people seemed especially onerous to the new director. He complained to Bolley that "in spite of this great decrease in the revenues of the Experiment Station we have made practically no reduction in the wages paid . . . . Either we must curtail some of our projects or reduce expenses of operation to keep within the reduced income." Trowbridge addressed the problem of technicians' wages in an imaginative way. Noting that "we consider ourselves fortunate if we have the means to have one high-priced man to head our investigations," he emphasized getting good student assistants who could aid the top scientists while they matured as researchers themselves. "It seems to me that is the only way we can build up our research work and train better workers at our smaller agricultural colleges," he concluded.[10]

Getting even a handful of quality scientists was not an easy matter. Under Trowbridge the Station began emphasizing systematically two means of recruitment that have endured. First, it leaned increasingly on native North Dakotans, researchers who had been at least partially trained at Fargo. Many of these people were already familiar with the work of the Station and were committed to it and the College. They were also usually inclined to sacrifice some potential income to live and work in familiar surroundings, close to family and friends. Second, the Station emphasized the hiring of young scientists from elsewhere with unrealized potential and sometimes without terminal degrees. This could be a problem, because promising young people learned on the job at Station expense. Chris Jensen remembered that when he came to Fargo in 1926 "I was completely untrained in research work," but he became the dairy products specialist and eventually fulfilled his promise.[11] Not everybody did. As a baseball manager supposedly said once, "potential is a French word for ain't done a damned thing yet." Sometimes people failed to reach their potential but they stayed on. Sometimes they reached their potential and moved on. Those were the risks inherent in this necessary policy. Trowbridge nurtured those who stayed and met his expectations, raising their salaries and sending them back to school for more training if they needed it. He understood what had to be done to maximize the quality of a small institution with few resources.

Trowbridge's modification of the Cooper structure was a necessary response to a difficult financial situation. It did not signal his intention to reverse the course set by his controversial predecessor. Indeed, in most areas Trowbridge embraced the tenets of Cooperism.

That the new director was unwilling to reconceptualize the NDAES and its purposes should come as no surprise. Trowbridge understood the currents of publicly sponsored agricultural research, and he showed no inclination to swim against them. He believed in academic specialization, scientific rigor, central coordination of research efforts, and clearly defined, tightly drawn plans of work. He did not believe stations should regulate or that researchers should be advocates, especially in areas outside their expertise. Cooper was gone, but Trowbridge did not intend to return to the freewheeling and exciting times of John Henry Worst.

There were many signs that Trowbridge intended to complete the transition Cooper had initiated. He reorganized departments to reflect the greater disciplinary complexity of agricultural science, a process that had started before his arrival. Agronomy, Animal Husbandry, Dairy, Cereal Technology, Entomology, and Agricultural Engineering all were formed by the mid-twenties. He stressed specialization, rigor, and coordination of research efforts. The rapid rise of the clever and dynamic agronomist H. L. Walster, who came to the Station in 1919 and a mere five years later replaced old-hand C. B. Waldron as dean of Agriculture, was an indication of the new spirit.

The creation of specialized departments of Farm Management and Marketing and Rural Organizations at the NDAES was another sign of the enduring impact of Thomas Cooper. Advocates of improved farm management continued to stress that farmers were largely the masters of their own fates. "Efficient methods leading to the highest net returns are as important in this industry as in any other line of business," argued farm management specialists Rex Willard and O. M. Fuller in 1927. "Those who are not in a position to accept new and better methods will undoubtedly either be forced out of business or be obliged to accept low incomes and consequently a very low standard of living." With the encouragement and support of John Shepperd, agricultural economist Cap Miller pulled business-minded farmers together in the Northwest Farm Managers Association. The NFMA endures today as a supportive client group for the Station and an effective advocate of improved agricultural and business practices in farming.[12]

The pains some Station researchers took during the twenties to disassociate themselves from the pre-Cooper Station, its prominent public positions, and its leading figures provide more indications of the triumph of Cooperism. In 1924, for example, cerealist C. E. Mangels refuted the gospel according to Edwin Ladd, contending that "the durum wheats as a class are distinctly inferior to hard red spring wheats in bread making quality." Marketing specialist Alva Benton slapped at Ladd and others who had struggled successfully for fairer grain grading standards by arguing that "the real solution to the dockage problem lies in improved soil and crop management and in better farm practices." And researcher Rex Willard blamed struggling farmers for their own problems, contended that failure resulted from "wrong selection of enterprises, improper organization of the farm, inefficient operation, lack of capital, [or] poor land."[13]

The times had clearly changed, and yesterday's fashions were now out of date. Still, there was a mean spirit to remarks such as these that was—and is—uncharacteristic of the NDAES. Those who took pleasure in attacking the people who had built the institution demonstrated little nobility. Those who blithely blamed failing farmers for their own problems showed little compassion. The public display of such attitudes was also politically unwise. Many North Dakotans still revered the early Station giants and appreciated their services to the state. And many farmers on the edge of failure did not believe the blame was mainly theirs. The Station needed to build support in all segments of the population, and cruel, arrogant, and insensitive comments did not advance that end.

It should come as no surprise that in the conservative environment of the twenties the sort of public advocacy characteristic of the NDAES during the progressive period disappeared completely. The only Station publication remotely resembling political advocacy during the twenties was a puff piece by Benton and Fuller for the proposed St. Lawrence Seaway. That research was commissioned by the Greater North Dakota Association. The Station had traveled far from the days when it battled for the common people against the economic powers that dominated the state.[14]

The Station's embrace of Cooperism aroused surprisingly little internal controversy. Ladd resigned the presidency in 1921 to take a seat in the United States Senate. His successor, agricultural economist John Lee Coulter, strongly supported efforts to make the NDAES a conventional and conservative research institution. Henry Bolley resisted, albeit rather halfheartedly. He had spent himself fighting Cooper and lacked the energy for another full-scale effort. He did make a strong public case for activism, arguing for a rededication of the land-grant colleges to the service of the people. Bolley's problem was that his colleagues were moving with the current of the times away from broad public activism and responsibility.

That is not to say that idealism was dead at the NDAES, but it was different. The younger generation hoped to contribute to human betterment through their academic specialties; they did not feel the obligation to redeem humankind from all of its problems. They tended to be more narrowly technical, and as specialists, not broadly trained generalists, they were reluctant to speak with authority outside of their limited areas of expertise. They also were inclined to be careerist. Ted Stoa became an agronomist because "most of the service you could do for the people of the state was in agronomy," but also because there were "Federal jobs" in that field. Chris Jensen was drawn to dairy work because it was "a service to mankind," but also because "I thought, here is something that will be stable." For Stoa and Jensen and their generation, agricultural research was more of a job than an adventure—they might have listened to Bolley, but they did not really hear him.[15]

Molding Bolley to the new realities commanded much of Trowbridge's attention during his tenure. In 1919 he held a meeting with Ladd, C. B. Waldron, and Bolley designed to induce the botanist to stop doing the work of Agronomy. But the next year he complained to Bolley that "there has been a very great tendency on the part of your department to spread itself over the entire field of plant life. The Station policy has been recognized as one of departmental divisions with departmental responsibilities." He pointedly reminded Bolley that "we have not the means of maintaining two agronomy departments." Bolley surrendered his wheat work. The next year there were problems with Bolley's role as seed commissioner. Trowbridge believed he had been careless about releasing some seed in his charge.[16]

Regulatory work was a sore point with Trowbridge anyway. Ladd's seat was hardly cool before the food commissioner's job was sluffed off to Bismarck, and Trowbridge almost certainly looked forward to being rid of seed regulation as well. Eventually he got his chance. In 1927 the legislature amended the seed law, giving the Board of Administration authority to appoint the seed commissioner. Previously the law had designated the Station botanist as commissioner. The following year there were major damage problems with certified seed potatoes sold in Southern markets. Bolley was probably not to blame. Seed potato certification was the responsibility of the growers. Still, Bolley was stigmatized, and in 1929 the Board of Administration was reluctant to reappoint him for another two-year term. Bolley wanted to be reappointed, and Coulter agreed to recommend him, though the president wrote the board that "we would be glad . . . to see Professor Bolley give all of his time to . . . the Experiment Station . . . because there is a great opportunity for Professor Bolley to serve through his research work." Trowbridge suggested that Bolley was "getting to an age" where seed regulation plus his regular work might be too much to handle. The board declined to reappoint Bolley, and the connection between the Station and that regulatory function was severed. Bolley thought he lost the job he had created because of political retribution for his past activities. "You . . . may not be aware of the fact that I am no longer State Seed Commissioner," he wrote his old friend Worst in 1930. "My 'good friends' in and about this institution became afraid that I would work myself to death in . . .

that job."[17]

Bolley's ironic comment on his "good friends" was apt. Rodney Hastings, whose career in seed certification began under Bolley, noted that "by 1929 Bolley was old and kind of flighty . . . . He wasn't too popular amongst the other scientists—he was kind of unorthodox." Bolley lived on, confining his attention to flax. Indeed, he lived long enough, and became harmless enough, to become a sort of beloved anachronism at the NDAES.[18]

The separation of the Seed Department from the NDAES marked the culmination of an endeavor launched by Thomas Cooper fifteen years before. Withdrawal from regulatory responsibilities made the Station a more conventional scientific institution and allowed researchers to concentrate more fully on what they were trained to do. For these reasons Trowbridge and his colleagues undoubtedly welcomed the end of the Station's significant regulatory role. The problem was that regulation could not be seen simply in scientific terms. Regulatory work had allowed the Station to develop a favorable public image and generate public support. It had allowed the NDAES to show the flag everywhere in the state and to make a difference in the lives of most North Dakotans. Worst, the politician, realized the utility of regulatory work; Trowbridge, the scientist, did not.

When the Station became a more conventional scientific institution its image in the agricultural research community improved. But Trowbridge forgot that North Dakotans pay the bills. Scientific reputation was not very important to them, especially because nobody at the Station bothered to inform them why it should be. The time would come when the NDAES would need the favorable public image and the public support that regulatory work had provided. Only then would it realize the negative implications of a step that had seemed so positive in 1929.

II

The painful and controversial withdrawal of the NDAES from regulation obscured the basic continuity of most of its research work. As before, the Station focused most of its attention during the twenties on the stubborn production problems faced by North Dakota farmers. Developing appropriate crops and methods for a difficult environment as means of gaining greater economic security for farmers and the state remained the predominant theme at the NDAES.

No area of the state presented greater production challenges than the Missouri Slope in the west. To the dismay of some boosters of that region the NDAES remained wedded to the vision of animal rather than crop production there.

If anything, the notion that the Slope should mainly be range was more prominent at the Station during the twenties than before, now that the euphoria of the boom had passed away and the USDA had ceased to fund most of the dry land experimentation it had been doing in cooperation with the NDAES. The USDA lost faith in the possibilities of dry land agriculture in part because they seemed so meagre. It was also the case that the commodity glut of the early twenties indicated that the nation no longer needed to explore the productive possibilities of regions perceived to be marginal.

The NDAES shared the USDA's disappointment with the possibilities for dry land agriculture. The superintendent of the Williston substation argued in 1922 that "no known method or 'system' of farming can produce satisfactory crops in the often recurring dry years. And . . . grain crops are too uncertain and are grown at a loss in too many years to be dependable, and . . . if people are to remain in this or similar sections, . . . make a living and get a fair return from the land, the annual farm income must be stabilized by diversified farming with feed crops, live stock and live stock products as the main industries." The Station continued to develop appropriate crop varieties for the western part of the state, thereby enhancing its agricultural potential. The early NDAES suggestion that Slope producers were best advised to engage in a mixed agricultural system in which livestock played a prominent role, however, has generally proved to be wise.[19]

The NDAES continued to stress the importance of diversification for income security, not only in the western portion of the state but in every other section as well. North Dakota farmers made some progress in reducing their dependence on wheat during the twenties. In 1920 over half of the cultivated acreage in the state was seeded to hard red spring or durum wheats, but in 1930 only 43 percent of the planted acreage was in wheat. In the same period flax and barley acreage both more than doubled and the amount of land devoted to corn rose by 44 percent. The acreage seeded to tame hay increased by 63 percent, an indication of farmers' growing interest in livestock.

The Station had always stressed the utility of dairy as a means of providing steady income and maintaining a balanced and efficient agricultural operation. There were signs during the twenties that farmers were listening.By 1923 dairy products had become a source of farm income second only to wheat. Between 1920 and 1930 the number of North Dakota milk cows rose by 32 percent, and farm income from the sale of milk and cream jumped by 52 per cent.[20]

When one looked beneath the impressive statistics, however, it was not as clear that the Station was winning the war for the dairy cow. Two investigators noted in 1930 that "dairying in North Dakota is still a side line business with most farmers." That reality was reflected in the relatively small number of cows per farm, in low production per cow, and in problems of poor quality caused by careless handling and tardy delivery of milk and cream.[21]

Other once-promising potential routes to diversification also proved disappointing. Accomplished horticulturist A. F. Yeager continued to develop appropriate fruit and vegetable varieties for North Dakota, but he admitted that the vagaries of climate and market dictated that most fruits and vegetables be grown for home use. In 1928 Alva Benton tempered the good news that North Dakota chickens and eggs were selling on the Chicago market with the disturbing report that most of the former were substandard and the latter were dirty. Climate, distance, and the inclinations of North Dakota farmers continued to frustrate those who hoped for a dramatically more diverse agriculture in the state.[22]

Wheat was still king in North Dakota, despite the relative rise of a few petty pretenders. Reflecting that fact, Agronomy became the king of departments at the NDAES during the Trowbridge years.

Agronomy developed as an independent discipline around the turn of the century and the Station employed its first trained agronomist, O. O. Churchill, in 1904. Churchill's hiring reflected the Station's appreciation of the importance of agronomy to North Dakota farmers and ranchers. Agronomists concentrated on perennial grasses, such as alfalfa, bromegrass, clover, and crested wheatgrass, as well as annual grains, such as barley, corn, oats, and wheat—the crops, in short, of greatest significance to North Dakota.

To fulfill its potential at the NDAES agronomy needed independence and aggressive leadership. President Ladd put it on the road to independence in 1916 when he gave it departmental status. Agronomy advanced its cause in the early twenties when it succeeded in wresting all of the grasses and annual grains but flax away from Bolley. In 1919 the department gained strong and confident leadership when Harlow L. Walster, a University of Chicago Ph.D., was brought in as head. A brilliant and widely-read scholar, Walster stressed rigorous scientific research and staff teamwork. In 1924 he became dean of Agriculture and a decade later he succeeded Trowbridge as director of the Station.

Walster's key researcher was wheat breeder Lawrence Root Waldron, who in 1896 had come to Fargo to be with his older brother, pioneer horticulturist C. B. Waldron. He graduated from the NDAC and went on for a masters' degree at the University of Michigan. After a stint with Bolley in Botany, Waldron went to Dickinson as superintendent when that substation opened in 1905. In 1916 the forty-year-old Waldron was brought back to the main station to assist in the search for a rust-resistant wheat. He completed his formal education in the late twenties when he received a Ph.D. degree from Cornell.

L. R. Waldron was a complex man. Introverted and bookish, he had a wide range of scholarly interests and could usually be found at the library when he was not in his lab or on his plots. His scientific abilities were unquestioned. USDA flax researcher Harold Flor considered him the strongest researcher on the staff. "Waldron was a real scientist," Flor contended, "the only one [at the NDAES] I would class as a first-class scientist." Unfortunately, he lacked patience with those not as intelligent as he, and assistants often were uncomfortable working for him. Eventually he came to believe his accomplishments had been insufficiently appreciated, and he ended his career a bitter and resentful man. This was regrettable, because his work benefited North Dakota greatly. Indeed, Glenn Smith has estimated that "the work of L. R. Waldron would more than pay for the whole cost of the College and Station for its entire history up to 1970."[23]

Waldron was complemented in Agronomy by Ted Stoa. A native North Dakotan and an NDAC graduate, Stoa was devoted to the state and its agriculture. Stoa was not a brilliant scientist, but he was careful and conservative, thorough and patient. A team player and a modest, self-effacing man, he never forgot that the Station existed solely to serve the people of the state. He was one of the scientists whose loyalty and hard work in helping the Station through the difficult years of the thirties were rewarded by strong public support in the post-World War II years.

During the twenties the agronomists worked on all sorts of crops—save Bolley's flax. Winter wheat commanded much of their attention. Its high yields, low vulnerability to rust, and minimal labor requirements were all attractive, but it was always a high-risk crop in an area where severe cold and light snow cover were common. Agronomy also labored at developing and promoting crested wheatgrass. An early starting, rapidly-growing perennial with a deep and intricate root system, crested wheatgrass was far superior to alfalfa or bromegrass as range for the western counties of the state. Its hardiness, drought resistance, and palatability made it a favorite on the Slope.[24]

Hard red spring wheat was the crop that commanded most of the agronomists' attention. In North Dakota the major enemy of spring wheat was stem rust, a parasitic fungus that periodically attacked the crop, often with devastating results. Ever since the Station's early days combatting stem rust had been a major enterprise. Durum had become popular largely because of its rust resistance. Bolley's correct belief that the barberry bush served as the host for rust spores helped trigger a federally supported barberry eradication campaign in the wheat belt. The agronomists wanted to go beyond mere prevention by systematically breeding rust-resistant strains of hard red spring wheat. Waldron was brought to Fargo during the rust outbreak of 1916 to do precisely that.

The immediacy of the rust problem, along with the element of creativity in the endeavor and the fact that, in the words of Glenn Smith, "it is flattering to see the state covered with your variety," made wheat breeding exciting and potentially satisfying work. But it was not easy work. For one thing, rust resistance was not enough. It was also necessary to have high yields and early maturation appropriate for North Dakota's short growing season. Wheats also had to produce flour that met the exacting standards of the baking industry. In 1922 Twin Cities millers complicated wheat breeding further when they began paying a premium for high protein content—or downgrading wheat with low protein content. This added yet another demanding variable. Wheats now had to be bred for "high yield, earliness, rust resistance, and high technical value." As Station veteran David Ebeltoft said, "It's like trying to get a brown-eyed blonde."[25]

The search for the brown-eyed blonde began to bear fruit in the twenties. In 1922 the Station introduced Kota, a durum-hard red spring cross that Waldron had discovered in 1917 and had developed in cooperation with the USDA. Kota met the criterion of resistance to stem rust, but it was vulnerable to smut and leaf rust and was an inconsistent yielder. Waldron then crossed Kota with Marquis—the variety of hard red spring wheat favored by farmers and millers alike—and produced Ceres. Released in 1926, Ceres was rust resistant and high in protein and baking quality. It was also a dependable yielder of better-than-average crops. Eventually it covered over 40 percent of the spring wheat acreage in North Dakota.[26]

Kota had been valuable as a parent for Ceres, but its release had stimulated a conflict between the millers and the NDAES. The millers complained bitterly of what they saw as low quality in Kota, and they protested the Station's release of this new variety without prior consultation with them. This criticism stung the Station and was taken quite seriously because, after all, it was the millers who purchased the wheat from the farmers.

In order to prevent a recurrence of the Kota controversy the Station decided to "permit industry to share in the evaluation of . . . new lines." It did so by furnishing millers with "samples of new . . . varieties . . . which they might test. In this way industry could become acquainted with a new line before it was released or recommended, and actually might aid the experiment station in the appraisal of the new line." Agronomy soon began serving as a go-between, contracting with farmers to produce experimental varieties in commercial quantities for the millers' tests. This arrangement was formalized in 1928 with the creation of the Northwest Crop Improvement Association. This new organization was made up mainly of businesses in the grain trade, "such as mills, line elevators, terminal elevators, grain commission companies, independent and farmer elevators, malting firms, and banks and railroads." Its stated purpose was "to improve the quality and increase the yield per acre of all cereal crops in areas tributary to Minneapolis and Duluth terminal markets." The Station frequently and sometimes vigorously disputed the results of the NCIA's milling and baking studies. But agronomists and cereal technologists worked closely with the industry group and with its successors.[27]

This developing relationship reflected Trowbridge's stress on cooperation with agribusiness and was another sign that Cooperism outlasted Cooper. The Station had changed dramatically since the days when Ladd and the millers stood at swords' points. To the agronomists, who wanted to help the farmers grow quality varieties and receive high prices, cooperation with the millers made perfect sense. As Ted Stoa noted, "new and better lines must be developed and, when grown, must have a market." As long as the United States remained a country with a capitalist economy, that market would be dominated by private corporations. The problem with this relationship, which the NDAES could not foresee, was that it invited criticism from resurgent agrarian radicals during the thirties. To farmers on the left it appeared that the people in the Better Farming Association had been expelled from the Station temple only to return in the guise of the Northwest Crop Improvement Association.[28]

III

The orientation of the agricultural research system toward solving production problems raised criticisms during the twenties. After nearly a generation of enjoying ample markets for their products, farmers rather suddenly found themselves suffering from overproduction. Some organized farmers, politicians, and other public commentators on agriculture wondered aloud whether research resulting in increased production was helpful or even justifiable in a country awash with excess wheat, corn, pork, and cotton. A few, ignoring the effects of European agricultural recovery and a strong dollar on exports, even blamed the stations and the USDA for causing the surplus problem. The research system had been criticized before for its production orientation, but such criticism became much more intense during the agricultural depression.

Spokesmen for agricultural research attempted to parry these criticisms as best they could. They argued that overproduction was preferable to want or that it was just a temporary condition, and contended it was due in part to international economic developments for which they were not responsible. They stressed that research benefitted farmers by making them more efficient and lowering the cost of production. And they derided those who, in effect, argued that farmers were better off ignorant and unscientific. "Those in agriculture . . . who have the most knowledge concerning their labors are the best off," contended President Raymond A. Pearson of Iowa State College in 1925, "whether prices are high or low." Others maintained that abundant production benefited society, justifying the social resources that had been devoted to research. As one noted, "increasing the production for the farmer has oftentimes resulted in immediate financial injury, through low prices for his product, but it has always worked to the benefit of the consumer in lowering his food costs." It is noteworthy that spokesmen for research willingly took credit for increased production, even though agriculture was suffering and they were not able to point to any single development—such as hybrid corn, DDT, or anhydrous ammonia later—that had a major impact on agriculture. To do otherwise, however, would raise doubts about whether public expenditures on agricultural research were justified in the first place.[29]

The research system was rebuked during the twenties for what it failed to do as well as for what it did. Critics argued that problems of marketing and distribution, rural society, and family life had been slighted by stations concentrating single-mindedly on increasing production. Spokesmen for the system generally agreed with this assessment, and in a brilliant display of bureaucratic survival instincts, they requested money to correct the imbalance! The result was the Purnell Act of 1925, providing funds to the stations for the study of topics in agricultural economics, marketing, rural sociology, and home economics.

All of the stations readily accepted the Purnell funds because the appropriation allowed them to expand and because they saw themselves as the appropriate entities for undertaking all research dealing with agriculture or rural life. But they did not consider the legislation an unmixed blessing. Reviewing the first year of Purnell Act funding, Chief E. W. Allen of the Office of Experiment Stations noted that quality researchers in the new fields were scarce and that much of the early work had been superficial and amateurish. These same criticisms had been leveled against the stations in their early years and probably came inevitably to any new enterprise. More serious was the charge that because some states cut their appropriations when Purnell funds arrived the new emphasis would harm the traditional areas of station research.[30]

Underlying these complaints was the basic problem that natural scientists had difficulty accepting the legitimacy of the social sciences. Natural scientists dealt with plants and animals. Because these materials were relatively simple, stable, and manipulable, natural scientists were able to use them to prove hypotheses, replicate experiments, and verify findings. In other words, natural scientists could reach conclusions that were widely accepted through methods on which there was universal agreement.

Social scientists dealt with people and their attitudes, institutions, and behaviors. People were infinitely more complex than plants and animals because they were independent acting and thinking entities rather than mere collections of cells. Hypotheses that touched on humans outside the biological realm could be supported but seldom proved; people resisted manipulation and Western moral values limited the degree to which it could be attempted; experiments could seldom be replicated exactly and findings could never be completely verified. For all of these reasons the range of reputable theories and acceptable methods was much wider in the social than in the natural sciences.

R. W. Thatcher, the respected former director of the Minnesota station, captured the nature of the division between the natural and social sciences in 1927 when he noted that "the methods of research study are different when one deals with the physical problems of the universe from those encountered when one deals with the effects of personality and of social groups and social organizations upon existing situations." Another observer added in 1929 that "we have not arrived at that stage in human advancement wherein social experiments can be conducted on a laboratory basis, under conditions of satisfactory controls." Many champions of the social sciences believed that stage would be reached and that sociologists and economists would then be accepted as equals by natural scientists. Those who value the dignity of the human personality are thankful we cannot yet do with people what we do with plants and animals.[31]

There is no reason to believe that natural scientists at the NDAES had any more respect for economists and sociologists than did their disciplinary brethren in other institutions. They already had some acquaintance with social scientific methods and practitioners, however, in part because agricultural economists had been involved in the Station's promotion of better farm management. Moreover, the leading social scientists at the Station during the twenties had come before the Purnell Act was passed. Their research agendas were already largely set and were simply expanded with the federal money. Rex Willard and O. M. Fuller, for example, were already following the economic direction of North Dakota agriculture as part of their farm management work. It was not a happy story. Willard reported increasing tenancy, low commodity prices, high fixed costs, and deteriorating farm living standards. It was news to nobody in North Dakota that agriculture was in serious economic shape, but Willard helped make it clear how difficult the situation was.[32]

A favorite solution for farmers' economic troubles during the twenties was cooperative marketing. In 1921 Congress exempted co-ops from the antitrust laws with the Capper-Volstead Act. This legislation helped stimulate the American Farm Bureau Federation to launch an impressive promotional effort for cooperation. The Federal Farm Board, President Herbert Hoover's answer to the agricultural problem, closed out the decade lending money to get co-ops started. In North Dakota the Farm Bureau and the Farmers Union both vigorously advocated cooperation, and the ambitious but ultimately unsuccessful North Dakota Wheat Growers Association attempted to market the state's leading crop cooperatively.

NDAES economist Alva Benton and several colleagues used Purnell funds to study the practice and possibilities of economic cooperation in North Dakota. Station support for cooperation might appear to be a major exception to the rule that Worst-era advocacy was jettisoned at the NDAES during the Trowbridge years. It is important to remember, however, that in the twenties cooperation was embraced as a panacea by most of the agricultural establishment. Clearly, nothing endorsed by Herbert Hoover could be all that radical.[33]

Purnell money allowed a dramatic expansion of the work of rural sociologist E. A. Willson, who was attached solely to the College before the new funds came in. Willson's major interest was rural social institutions and the impact of economic changes thereon. As he surveyed the institutional environment of rural North Dakota, Willson argued that "from the standpoint of providing 'a good kind of life on the farm' many of the social organizations and agencies . . . are inadequate." He pointed especially to poor health care, insufficient recreational opportunities, weak churches, and inadequate schools as problems. Willson carried into this new field the proud Station tradition of telling North Dakotans the truth, whether they wanted to hear it or not.[34]

The Purnell Act also allowed the Station to fund projects in home economics. In the late twenties these mainly involved nutritional research on meats conducted by Esther Latzke and Constance Leeby.

As a result mainly of federal largesse, the Station was involved in more different areas in 1929 than ever before. These were probably not areas in which the NDAES would have chosen to be had it enjoyed complete discretion. But in retrospect it is clear that the rise of social science and home economics at the Station allowed it to serve the state more fully and touch the lives of more people. This broadening of the Station's appeal compensated partially for the diminution of its activities as a regulator and advocate of public policies.

IV

In the late twenties the NDAES appeared to be emerging from the long and painful period of readjustment that had started more than a decade before.P. F. Trowbridge had been at Fargo long enough to put his ideas about what a station should be into effect. He had gathered a nucleus of young and promising researchers, who provided continuity after the disruptive personnel changes in the Cooper period. The Purnell Act provided increased funding and allowed research in hitherto unemphasized fields. Researchers continued to make solid contributions in traditional areas of interest, and successful wheat breeding raised the prospect of dramatically greater service to the economy of the state.

Conditions seemed to be improving in the state as well. Agricultural prices remained unsatisfactory to most producers, but greater stability followed the failures of marginal farmers, bankers, and businessmen in the early and mid-twenties. A near-record wheat harvest of over 155 million bushels in 1928 brightened the farm income situation considerably, if temporarily. The bitter political divisions of the teens generally receded below the surface, imparting a greater quiescence to public life.

All of this enhanced stability for the Station. The decline of political turmoil and the more conservative public temper allowed Trowbridge to direct the NDAES into the scientific mainstream and establish a working relationship with agribusiness with little criticism. Bolley's dismissal as seed commissioner and the divorce of that office from the NDAES passed with barely a murmur in the state. The Station's lower political profile and the state's improved economic situation translated into increased financial support for it. In 1927 the legislature raised appropriations for the Station to more than $307,000, an increase of nearly $42,000 over the previous biennium. This added income allowed substation work to resume at pre-1921 levels and made it possible to reconvert the Williston substation from a county poor farm to an agricultural research facility. In the next biennium state appropriations dipped to $303,000, but federal contributions reached a new high of $150,000 and income from sales and fees totaled $178,000. This support was sufficient to maintain a staff of fifty-nine, including five USDA employees. The Station had again reached the staffing level it had achieved in 1914.[35]

Unfortunately, the late twenties was a false spring for North Dakota and for the NDAES. Soon the state was plunged into economic and environmental disaster. The Great Depression and the Dust Bowl conspired to bring the state unprecedented hardship and to raise doubts whether it would even survive as a viable entity. As the state went, so went the Station. Appropriations plummeted, wages were cut drastically, positions went unfilled, work was slashed back, and morale fell. Added to economic stringency was a political assault that resulted in the dismissal, demotion, or resignation of several key scientists. The twenties ended on a hopeful note, but the NDAES soon faced the darkest and most trying days of its short and turbulent history.

6

HITTING BOTTOM

The Great Plains might well be defined as an area of stress and strain: they have been and probably always will be the stage upon which the American people will try their great social and economic experiments. They will continue to be the stage upon which Mother Nature will reveal herself in the greatest variety of moods, ranging from sumptuous plenty in the bountiful years to niggardly scarcity in the years of drought, hot winds, grasshoppers, rust, or other disaster.

H. L. Walster, 1936

It was a tough time, a discouraging time . . . . Morale was low, absolutely.

Ted Stoa, 1984

The false spring for North Dakota and its agricultural experiment station during the late twenties was followed by a winter of despair. In the fall of 1929 the stock market crashed. Distress among investors was soon felt by financial institutions and industry. By mid-1932 industrial production stood at barely half of 1929 levels, gross farm income was 40 percent of what it had been three years before, and perhaps one fourth of the work force was unemployed. The nation was mired in the worst economic depression in its history.

North Dakota was far from Wall Street, physically and spiritually, but it suffered intensely from the depression that was triggered there. The relative decline in agriculture was greater than that in industry during the depression, and North Dakota was an overwhelmingly agricultural state. Wheat was hit harder than most crops, and North Dakota was a wheat state. Capital dried up, and North Dakota was capital poor even in good times.

The economic disaster would have been bad enough, but Mother Nature, in the image crafted by H.L. Walster, chose to play a tragedy on the North Dakota stage. Nine of the eleven years between 1929 and 1939 were dry, some desperately so. As a result North Dakota participated fully in the Dust Bowl popularly associated with the Southern Plains. By 1934 the situation was grave, especially in the Drift Prairie and Missouri Slope regions. In that spring frightening "dusters" moved across the horizon as the wind lifted layers of topsoil off the plowed fields. Dust drifted across roads and into ditches and sheltered areas. Fences, farm implements, and even small buildings were buried. Motorists used their headlights on sunny days, and towns turned on streetlamps to cut the brown gloom that hovered suspended between the tortured earth and the bright skies. The dusters looked like blizzards but they were worse. The dust sifted into buildings through the tiniest cracks, covering everything with a fine but gritty film. Animals sometimes suffocated, the insides of their lungs literally coated with mud. People wore surgical masks, but still they coughed and sneezed. Children and the elderly contracted dust pneumonia, and some of the weaker ones died. It was a situation of cruel and exquisite irony. North Dakota was literally choking to death on the means of its livelihood.[1]

With the dust came grasshoppers, in seemingly greater numbers every year. What the dust failed to whither and the wind failed to blow, they ate. Desperate to control them, farmers sprinkled bran laced with arsenic on the ground. Well into the twenty-first century the water supplies of some towns will reflect that decision in their high levels of arsenic contamination. In 1935 things got a bit better, and it looked as if the state might make a decent crop. Then a virulent strain of rust hit, and Ceres, the supposedly rust-resistant variety that covered 40 percent of the state's spring wheat acreage, proved especially vulnerable. The next year the drought returned, worse than ever.

One can gain a sense of the dimensions of this agricultural disaster from statistics compiled by the commissioner of agriculture. In the last three years of the twenties North Dakota farmers harvested an average of 72.9 million bushels of hard red spring and 56.7 million bushels of durum wheat. From 1934 through 1936 hard red spring production averaged 21.6 million bushels, while an average of 10.1 million bushels of durum was produced. New Deal acreage limitation programs had a negligible effect. Between 1934 and 1936 10.3 million acres was seeded to wheat, down only 400,000 acres from the 1927-29. The big problem was yields. Hard red spring yields averaged only 5.2 bushels per acre on the land farmers bothered to harvest in 1936. Walsh and Pembina were the only counties in which the yield exceeded ten bushels per acre. In 35 of the state's 53 counties the average yield was 3 or fewer bushels.[2]

Figures such as these provide the statistical representation of a human problem. North Dakotans struggled against the curses of depression and drought, but each consumed the means needed to fight the other. The result was economic hardship and human misery unprecedented in the state. Land values plummeted. Tax delinquency became endemic. Unable to meet mortgage payments or pay property taxes, farmers lost land to banks and to government. The Bank of North Dakota, the state agency that was supposed to lead the farmers out of thralldom to private financial institutions, found itself in the embarrassing position of owning huge amounts of foreclosed land. At least one in three farms in the state was foreclosed, and forced sales became the most common way of transferring land. It would have been worse had not many governmental entities been lenient and many bankers seen the futility of liquidating virtually worthless land.

The farmers did what they could to hold on because they had no place to go. They became more self-sufficient, raising hogs, dairy cattle, and poultry,and generally "living at home." Onetime owners became tenants, sometimes on their own farms. Households grew as unemployed adult relatives came back from town to keep body and soul together and as older children, lacking alternatives, stayed on the farmstead. Try as they might, all could not stay. They streamed away in search of work, often to the Pacific Northwest, driven by dust and depression. North Dakota's population hemorrhage, so evident in the twenties, resumed. Had John Steinbeck lived in Washington rather than California, *The Grapes of Wrath* might have been about Olsons rather than Joads.

Those who depended on the farmers, as virtually everybody did, suffered along with them. It was a struggle to maintain churches and schools or continue even minimal governmental services. The market towns, founded on bright hopes of local and state greatness, suffocated, their false storefronts and pretentious brick business blocks mocking reminders of the inflated dreams of years past. The cities were only slightly better off. On Christmas Eve of 1936 philanthropic Fargoans quickly exhausted the five hundred food baskets they distributed to the needy at the Avalon Ballroom. The centerpiece of each package was a single chicken.

The desperate overcame their pride and their belief in independence and self-sufficiency. They looked for help wherever they could find it. State government did what it could. The imaginative and energetic William Langer, elected governor in 1932, cast about for solutions. He cut state spending in half, declared a moratorium on farm foreclosures, and embargoed the shipment of wheat out of state until prices rose. The moratorium further crippled local government and drove potential private lenders away, and the embargo had little tangible effect. Moreover, both actions were probably unconstitutional. But at least Langer was doing something in this time of desperation.

Supplicants even begged the North Dakota Agricultural Experiment Station for help—and not with better crops or improved bookkeeping, because it was too late for that. People pleaded for immediate economic aid. "I was foreclosed on and my land sold," a Northwood farmer wrote Henry Bolley in a letter requesting a loan. "I have lived here 50 years on [the] same farm." Even Treadwell Twichell, that conservative symbol of self-sufficiency, wrote his friend John Shepperd that he might need a job at the Station because he was on the verge of losing his farm. Shepperd had the courtesy not to advise Twichell to go home and slop the hogs.[3]

The Station was not in a position to provide financial help to others in the thirties. For a time it was an open question whether it would even survive as a viable entity. State appropriations were virtually eliminated, income from fees and sales of products shrank, and the future of federal appropriations was in doubt. The NDAES was not the only experiment station to suffer financial stringency, but its hardship was much worse in degree. It had been closer to the margin than most when the depression began, and it was cut more than most during the course of the thirties.

The result was dramatic retrenchment. Salaries were slashed, funds for equipment were reduced virtually to nothing, and researchers who left were not replaced. Everything declined but the demand for Station services. Indeed, the dismal duo of drought and depression spawned a number of new agricultural and human problems that the Station struggled admirably to address. But it was not possible to do everything. Secondary lines of work were abandoned and only the barest essentials were funded. As always, the NDAES did its best with what it had, but it had so little that its best could not be very good. "It was a tough time, a discouraging time," agronomist Ted Stoa remembered. "Morale was low, absolutely."[4]

The state and its people needed help to survive the thirties. Their actions were courageous and dramatic, but insufficient to overcome the crisis. After 1933 North Dakota leaned increasingly heavily on the federal government. Out of Franklin D. Roosevelt's New Deal administration flowed a stream of action programs. Some refinanced farm mortgages. Others used crop loans to put a floor under agricultural prices and rehabilitation loans to help dusted-out farmers try again. Tenants received money to buy land, and farm organizations got loans to buy cooperative facilities. Farmers were paid to not plant crops and to conserve highly erodible acreage. Shelter belts were planted to reduce wind erosion, and soil-conserving techniques were demonstrated. Farmer-owned cooperatives received loans to electrify rural homes. Food stamps were distributed, and work relief was provided for the unemployed. In 1936 half of the people in North Dakota were dependent on relief and many of the rest received other federal transfer payments.[5]

The NDAES had no staff members on relief, but without the federal money it received the Station would have ceased to exist as a scientific institution. Sixty-five percent of Station income in the 1933-35 biennium came from the United States government.

The federal funds for state and Station were a mixed blessing. Bureaucratic strings were attached, stimulating much irritation. And the money was never sufficient to do much more than keep the state on its feet. North Dakotans complained that they did not get their due, but what they got helped immensely. Perhaps the state would have survived as a viable social, political, and economic entity without federal help, but that help made survival a certainty.

The clouds of the thirties had a silver lining. Farm prices afterward were more stable as a result of federal programs. Federal loan programs made property more secure. Farmers emerged more conscious of their practices and more careful with their soil. The Farmers Union used federal loans to build the sort of comprehensive cooperative structure of which North Dakota farm groups had long dreamed. Electricity transformed the rural home and wiped out the biggest single difference between rural and urban life.

There was a silver lining for the Station, too. After the depression North Dakota farmers seemed more appreciative of what science could do for them. More of them cherished the Station and its activities on their behalf. They seemed to appreciate the scientists and the loyalty of those people to the state. For its part, the Station relearned the importance of serving public needs and of maintaining a favorable public image. It developed a congenial relationship with the major political groups in the state, from the Farmers Union on the left to the Greater North Dakota Association on the right. As the thirties receded into the past the NDAES increasingly found itself where it had always wanted to be. Getting there was the hard part.

I

During the early years of the depression there was some doubt, in North Dakota and everywhere else, whether the enterprise of publicly sponsored agricultural research would survive the economic calamity. Part of the problem lay in the fact that government at all levels, ground between rising demands and shrinking resources, slashed all nonemergency spending mercilessly. But a more fundamental difficulty was that many believed agricultural science offered no solution to farmers' pressing problems. Indeed, agricultural science was commonly held responsible for causing the basic farm problem of overproduction. Arguing that agricultural research was irrelevant at best and harmful at worst, politicians in Washington and the state capitals raised a chorus against appropriations for research at the stations and the USDA. In 1933 the retrenchment-minded Lewis Douglas, budget director in the incoming Roosevelt administration, suggested ending federal funding for agricultural research. Had his advice been followed precious few stations would have survived. The NDAES would have been one of the casualties.

Agricultural scientists defended their endeavor as best they could, repeating some of the arguments they made in response to similar criticisms in the twenties. Hoover's secretary of agriculture, Arthur Hyde, maintained that agricultural research was concerned with more efficient production, not greater production. "Scientific methods lead to greater production per acre, at lower cost," he contended, "and do not necessarily result in more acres and greater volume." Alluding to domestic hunger and restrictive European import policies, Andrew Boss of the Minnesota station concluded that "it is doubtful if there is serious overproduction at the present time, if means for economic distribution were provided and purchasing power maintained." Boss concluded that "abundant . . . production must ever be the objective of the successful farm operator." Other commentators suggested that production-enhancing research was a prerequisite for agricultural stability. "The fundamental principle of agricultural research is to give mankind greater and greater control over the forces of nature," argued T. O. Walton of Texas A and M. "The more nearly we approach our goal in harnessing these various forces, the more nearly will it be possible to control production on a national or world scale, or to establish a stabilized enterprise on the individual farm."[6]

Criticism of agricultural science did not stimulate defensiveness only. It also led to some healthy analysis of the effect of agricultural research on farmers, agriculture, and society as a whole. Dean C. B. Hutchison of the University of California believed that agricultural research had increased economic pressures on marginal farmers and that "in some cases agricultural production has been expanded by new methods . . . to the point where even the good farmer receives so little for his produce that he cannot pay his debts and taxes." Hutchison and others agreed that consumers got the most out of agricultural research. "It is the public—all the people—who are the chief beneficiaries from scientific studies relating to food," noted Raymond Pearson of Iowa State College in 1932. "Farmers do not possess trade secrets, and they do not amass fortunes by taking advantage of scientific work to reduce their cost of production or to protect their crops and animals from destruction. They pass on the advantage to the consumers."[7]

There was much in what such scientists as Hutchison and Pearson said about the effects of what they did, and social benefits provided a strong intellectual argument for continuing agricultural research. The public paid for research with its tax dollars so it should reasonably expect to benefit. The problem with this argument was political. Farmers formed the major client group for agricultural research. Society in general was not easily organized into a political constituency for scientific institutions, however much it might have benefited from what they did.

Supporters of agricultural research more commonly argued that the challenges of the depression should be met by more rather than less science. As in the early twenties, the social sciences were touted as pathfinders through the economic wilderness in which agriculture was lost. As T. O. Walton conceded to the Land-Grant College Convention in 1934, "the weakness of this civilization is the lag of economic and social adjustment to the sweeping changes introduced by science. Our western civilization is in the midst of a severe test now because man has failed to control himself while he has been controlling the forces of nature." The component parts of the system displayed their characteristic adroitness at shifting in response to the currents of public opinion in the agricultural community. In the early thirties "planning" and "readjustment" were the agricultural policy buzz words, eclipsing, if not totally obscuring, the "marketing," "cooperation," and "farm management" emphases of the bygone decade. The research system offered its aid in realizing the new goals it helped articulate. As A. F.

Woods, the onetime Minnesota director who was now director of scientific work for the USDA, put it in 1932, "quite generally in the Department and the stations emphasis has been shifted to projects that are expected to furnish the *basic facts* on which recommendations for readjustments in agricultural operation must be based."[8]

The flexibility and survival instincts of the public system helped save it during the thirties. It was also well served by its clients in commodity groups and in general farm organizations, of which the American Farm Bureau Federation was the most important. In addition, enough key policymakers could see the wisdom of the long-term commitment the system represented to ward off its opponents' most serious attacks. Heavy expenditures on research over the years had the paradoxical effect of making it difficult and even foolhardy to consider ending publicly sponsored agricultural science. To do so would be to admit past failure and foreclose future possibilities. The agricultural research enterprise also benefited from the aggressive championship of Roosevelt's secretary of agriculture, Henry A. Wallace.

This Wallace was the last of a string of three distinguished Henrys from Iowa. His grandfather, "Uncle Henry," founded the highly regarded agricultural periodical *Wallace's Farmer* and served on Theodore Roosevelt's Country Life Commission. His father, Henry C., served as secretary of agriculture in the uncongenial Harding-Coolidge administration. Henry A. was an accomplished scientist whose work in the twenties contributed materially to the development of hybrid corn. In 1928 he broke with his family's traditional Republicanism to endorse Al Smith, who was running for the presidency against his father's old cabinet antagonist Herbert Hoover. He then became an informal advisor on agricultural matters to Franklin D. Roosevelt. When Roosevelt was elected president in 1932 he named Wallace agriculture secretary.

Wallace might well be considered the greatest secretary of agriculture. He presided over the department when it was intimately involved in the process of revolutionizing the relationship between the federal government and the farmers. No other secretary could claim to know agriculture, in its many facets, any better than he. None had a fuller understanding of the place of agriculture in American life or a greater appreciation of its importance to the future of the world than Wallace. And none had more inspiring ideals or a superior ability to communicate those ideals to the public. He was one of the great agricultural statesmen of American history, and were it not for his inopportune political radicalism, climaxing in his ill-fated presidential candidacy on the Progressive ticket in 1948, he would be ranked at least on a par with such demigods as Justin Morrill and Seaman Knapp. He is just now beginning to receive due credit for his accomplishments and his vision.

Secretary Wallace served notice that science would not be slighted if he had anything to say about it. "We might just as well command the sun to stand still as to say that science should take a holiday," he wrote in 1934. "Science has turned scarcity into plenty. Merely because it has served us well is no reason why we should charge science with the responsibility for our failure to apportion production to need and to distribute the fruits of plenty equitably." Wallace derided those who believed agriculture was a victim of its own efficiency. "It is vain to propose letting inefficiency take over the job of reducing the volume of the surplus," he argued in his annual report to the president in 1933. "Man is not built that way; if he were, he would still be in the primeval mud. Putting a brake on science is not the solution." Inefficiency would reduce production, he conceded, but the cost of such a retrograde movement to the farmer and the threat it would present to the United States and to the world was so great that it could not be seriously entertained.[9]

Wallace's devotion to science did not blind him to the shortcomings of the research system. He agreed with many friendly critics that there was insufficient cooperation and coordination among the stations and between them and the USDA. His secretaryship saw a struggle to rectify this problem by means that were sometimes creative and helpful and sometimes artificial and cumbersome. He was also a strong believer in "pure" or "basic" science in the USDA tradition, though he never denigrated applied work. As Assistant Secretary M. L. Wilson noted, he favored support for "long-time fundamental projects which are . . . expensive and apt not to produce concrete results for years to come." In part as a result of his urging, Congress included money for research in the fundamental principles of plant and animal life in the grab-bag Bankhead-Jones Act of 1935. Wallace's emphasis on basic research was seemingly vindicated in the thirties when hybrid corn, the product of a generation of work by many scientists in many places, was widely adopted by farmers. The development of hybrid corn still stands as one of the most revolutionary accomplishments of the public research system.[10]

Wallace also perceived flaws in the thinking and behavior of traditional scientists. He believed that scientists were generally naive politically and that they were not conscious enough of the uses to which their work was sometimes put. "As long as they could get enough money to pursue their researches," he wrote, "why should they care how someone else handled the social and economic power derived from these?" He also contended that scientists were timid, in part because of the criticism directed at activists such as Bolley and Ladd who had fished in troubled waters: "Those who delved too deeply into social and economic problems got into trouble, and so many of the best scientists felt it was not good form to do things which to certain types of mentality seemed impractical and which might endanger the financial support of science." Wallace maintained that the answer to these problems was greater social consciousness among individual scientists and in the system as a whole. "Science . . . has been creating another world . . . that simply must be motivated by some conscious social purpose if civilization is to endure," he argued in 1934. "Science and engineering will destroy themselves and the civilization of which they are a part unless there is built up a consciousness which is as real and definite in meeting social problems as the engineer displays when he builds his bridge."[11]

The thirties was a congenial time for such thinking. Never before were vested economic interests more discredited or derogated. Never before were the human problems that issued from economic difficulties clearer. And never before was the public more eager for governmental solutions to social and economic malfunctions.

In national agricultural policy there was a dramatic shift in emphasis. Stimulation of production was downplayed, and the adjustment of production to demand was emphasized. In place of the economic individualism that traditionally characterized agriculture the New Deal stressed economic democracy. Farmers were now to determine their own fates by voluntarily limiting production and changing land use patterns, urged along by the financial and rhetorical resources of the federal government. The economic and ecological crises of agriculture, so graphically represented by the depression and the Dust Bowl, were to be cured by democratic community action, assisted and gently guided by Washington. All sorts of new agencies appeared—the Agricultural Adjustment Administration, the Soil Conservation Service, the Resettlement Administration, and the Farm Security Administration, to name a few—that reflected the new emphasis. The new agencies provided a congenial environment for agricultural economists and rural sociologists, as did some old ones, like the Bureau of Agricultural Economics, which became more important under the new regime. Researchers from more traditional fields—range and soil scientists and ecologists, for example—were also welcomed, especially if they were friendly to the ideals of planning and readjustment.

It was an exciting and promising time for many in agricultural policy, but the new people and new emphases were not universally welcomed. Planning, however democratic and voluntary it might be, was not always congenial to people devoted to the traditional canons of individualism. Many agricultural scientists deplored the relative de-emphasis of production research. Programs to benefit farm workers and tenants or to help farm organizations buy cooperative facilities were derided by conservatives as social engineering. People in the Dust Bowl saw in land use planning a plot to depopulate the their region. The Southern elite complained bitterly that social scientists meddled in their social, economic, and racial arrangements and that New Dealers threatened the progress of the great modern enclosure movement that was rapidly displacing hundreds of thousands of sharecroppers from the land.

There were complaints about New Deal agricultural programs in North Dakota and at the NDAES, too, especially regarding bureaucratic confusion and complexity, but these were relatively muted. North Dakotans knew that they needed the programs, and they took advantage of them to a degree unequaled in most states. They were less ideologically averse to government action, having undertaken a good deal of it themselves. The state's leading farm organization, the Farmers Union, strongly supported New Deal action programs and received tangible benefits in return. The state's congressional delegation backed most federal initiatives in agriculture, complaining only that they did not go far enough. By the early forties North Dakota found itself increasingly out of step with a national agriculture marching to a more conservative beat. Eventually most of the action programs were emasculated or destroyed, but in their glory days they helped save North Dakota.

II

The fortunes of the NDAES have always been tied to the condition of agriculture in North Dakota. As agriculture declined and struggled for survival in the thirties, so too did this institution that served it.

New leaders emerged at the Station during these years of trial. President John Lee Coulter resigned in 1929 and was replaced by John Shepperd. The presidency of the College was one of the two positions Shepperd had long coveted. His tenure in the job provided support for the theory that when God really wants to punish us he answers our prayers. Shepperd had to cope with difficulties of unprecedented magnitude. Upon assuming office he discovered a substantial budget deficit that had to be covered. Draconian budget cuts and massive salary reductions soon added to his woes, as did increasing political interference from Bismarck. He struggled to keep things together, displaying his characteristic kindness, compassion, and good humor in the process, but the situation wore him down. By 1937 he was requesting to be relieved so that he could return to research work in Animal Husbandry. The Board of Administration complied in July, and a year and a half later he was dead.

Perry Fox Trowbridge continued as director of the Station until 1934. Burdened by political sabotage, financial stringency, and failing health, he accepted appointment by the board as "Director Emeritus" at that time. His replacement was H. L. Walster, dean of Agriculture and temporary director of Extension. Walster was more responsible than any other individual for guiding the Station through the difficult thirties.

Walster was recognized as a remarkable person almost from the moment he arrived in Fargo in 1919. A Wisconsin native, he had degrees from that state's university as well as from Harvard and the University of Chicago. A mere five years after he came to head the Agronomy Department he was made dean of the College of Agriculture. In that position he insisted on rigorous scientific work and cooperation among and within departments. He continued to serve as dean in 1934 when he also assumed the jobs of Extension director and Station director. Walster thereby set the enduring pattern of having the same person hold the posts of Station director and Agriculture College dean simultaneously.

Energy that seemed unbounded was one of Walster's most memorable traits. In addition to holding three full-time jobs, he kept close track of everything that went on in the Station, interviewing all job candidates and reading all publications. Walster was a man of wide learning and impressive erudition. He "was a tremendous man—a scholar all the way down the line," as one former researcher remembers him. His learning was one of the things that made him imposing and a bit intimidating. Members of his staff commented on his blunt and gruff manner which often cowed subordinates. He was much in demand as a speaker, and he honored requests to visit communities throughout the state, crafting a different address for each group and occasion. His responsiveness was one sign of his political sagacity; another was his assiduous cultivation of every important farm and business group in North Dakota.[12]

Walster was not without his flaws. He considered himself the most accomplished person at the Station and expected others to concur in that judgment. He never hid his light under a bushel. As Chris Jensen said, "he was one of those people who liked to impress people." And he encouraged sycophancy, at least indirectly, leading many at the Station to believe that those who flattered him were unfairly rewarded.[13]

Despite his shortcomings, though, Walster was one of a handful of key people in the history of the Station. His importance derives from the critical time in which he served and the way in which he handled the challenges facing the Station. He provided calm and steady leadership. He was realistic about the situation in the Station and the state. And he always told the truth about both, although he also loved both and refused to despair about their future. Glenn Smith's conclusion that Walster was of crucial importance "as a builder of the institution" is unquestionably correct.[14]

The main problem with which Shepperd, Trowbridge, and Walster had to deal during the thirties was that of money, or, rather, the lack of money. Acting as if it intended to begin depression financing early, the legislature cut appropriations to the NDAES in 1929. The 1931 legislature increased funding, but as the depression deepened the fiscal situation became gloomier. In 1932 the North Dakota Taxpayers Association, a conservative group in which Fargo businessmen played a prominent role, secured voter approval of an initiated measure decreeing that no public employee—including the governor—could receive a salary exceeding $2,400 per year. Taking its cue from the voters, the 1933 legislature slashed appropriations to the Agricultural College by 59 percent. Appropriations to the Station were cut even more dramatically. In the 1931-33 biennium the NDAES had received nearly $233,000 from the state. For the following biennium a bit over $29,000 was appropriated. This meant that state appropriations were reduced by 88 percent from one budget period to the next.[15]

Apparently the legislature reasoned that the Station could be greatly reduced because it had other sources of revenue. It did enjoy federal appropriations that totaled $180,000 in both biennia, but some of these funds were supposed to be matched by state contributions. The USDA sympathized with the plight of the NDAES, but the possibility always existed it would cut funding because of the unwillingness of the state to fulfill its responsibilities. Throughout the thirties the governor and legislature were unwilling or unable to see that the acceptance of matching funds obligated the state to increase or at least maintain appropriations, not reduce them. This embarrassed the Station in its relations with the USDA and kept it in a marginal economic position.[16]

State appropriations was not the only category in which income declined. In the 1931-33 biennium the Station earned over $92,000, mainly from sales of surplus products. In 1933-35 income from this source dropped to about $65,000. It was further reduced later when complaints of competition by local producers led to a sharp curtailment of sales of Station products. The total tells a depressing story. In the 1931-33 biennium the NDAES had $505,000 to work with. In the next biennium it had only $274,000.[17]

The salary problem was the most disheartening aspect of a dismal situation. As Cap Miller noted in the late thirties, "even in normal times and before the depression salaries at NDAC were below salaries in many similar institutions in the nation." These insufficient salaries were cut approximately in half by the combined actions of the voters and the legislature. Shepperd's salary stood at $2,400, deans and directors received about $1,800, and many instructors and assistant professors got under $1,000 per year. As a USDA employee Glenn Smith was paid more than Ted Stoa, though the latter was chair of Agronomy. USDA flax researcher Harold Flor estimated his salary was twice that of Director Walster.

Only the most confirmed optimists could maintain their morale in this situation. The usually confident H. L. Bolley wrote to his old colleague T. D. Hinebauch that "the terrible conditions which these depressing times have brought upon the institution and upon our work . . . makes me feel at times that our early struggles seem to have been almost in vain." To a former student he added that "the professors . . . are cut to salaries less than what they started out with in 1890." John Henry Worst urged Shepperd to appeal to "college patriotism" and reminded him that while "having one's salary cut more than half is a severe blow," it was important to "remember that not less than ten million families are now eating the bread of charity, with no work to do that will enable them to live as they once did, by drawing a salary."[18]

There was much to what Worst wrote in that dreadful year of 1933. But the fact that their situations could conceivably be worse did not provide much comfort to scientists losing homes, cars, and furniture because they could no longer make payments. To make ends meet researchers took second jobs or served as consultants to private agribusiness firms. New Deal action programs were a blessing. Alva Benton, Chris Jensen, and P. J. Olson were among those who worked for the Agricultural Adjustment Administration. F. W. Christensen, Cap Miller, and Ted Stoa did Bureau of Agricultural Economics work. Herbert Hanson worked on soil conservation for the Interior Department. And Howard McColly, P. F. Trowbridge, and E. A. Willson were some of the Station people who worked for the Federal Emergency Relief Administration and the Civil Works Administration. There were two major problems with this very necessary activity. The first was that it took researchers away from their Station work. The second was that it opened for some of them—Benton and Willson, for example—federal careers that totally deprived the NDAES of their services.

Despite heroic efforts by individuals to stay and by the Station to keep them, many left in the depression years. Some found employment with the USDA or other stations, and a few went back to graduate school. One Soils man quit to run a Civilian Conservation Corps camp, and the sheep specialist left to farm. That one would find farming more remunerative in 1933 than working at the NDAES was eloquent testimony to the discouraging situation at the Station.[19]

Staff attrition and moonlighting obviously had a negative effect on the volume of work undertaken and the efficiency with which it was carried forward. Other factors also contributed to deterioration. There was virtually no money for equipment. "During the thirties we were down to ground level," Chris Jensen noted, adding that his equipment "budget was $100 per year." Horticulturist A. F. Yeager wrote to Governor Langer in 1933, "my desk, which was originally worth about $10.00, is now twenty years old; other fixtures are in proportion. Laboratory space, scientific equipment and greenhouse space are badly needed. Visitors from other schools marvel that anything is done with our meager facilities." Yeager made it clear that cost to the Station was cost to the state: "If it were possible to install lights immediately in the greenhouses we now have, much more progress in breeding work could be made. Lights costing less than $1,000 would make another generation of plants per year possible, thus speeding up the work by 50 percent." Unfortunately, $1,000 was hard to come by in 1933.[20]

The Station struggled to adjust to the funding crisis. Researchers who left were not usually replaced. Hourly workers were let go, forcing scientists to spend time on physical labor and to depend on sporadic help by work relief recipients covered under the Civil Works Administration and Works Progress Administration programs. The Station interpreted the Purnell Act in such a way as to allow the shifting of funds to production-oriented departments. As a consequence, research in agricultural economics, rural sociology, and home economics periodically ceased. In home economics funded research was not resumed until the 1970s. Among the substations Dickinson alone operated at anything approaching the normal level. At each of the other locations the plots were seeded to grass, clover, or alfalfa, and the staff was reduced to one caretaker to look after the buildings. Even relatively favored departments such as Agronomy had to cut back. Stoa and Walster reported in 1935 that "variety trials with oats, peas, soybeans, buckwheat, mustard and sugar beets" had been discontinued. In 1933, the Station could not even afford to produce a biennial report. The calamities of nature compounded the Station's economic woes. Drought made it difficult to carry out field work and grasshoppers chewed up experiments.[21]

Adding to the Station's burdens in the thirties were new demands related to depression and drought. Ever conscious of its responsibility to serve the state, the NDAES attempted to address these to the best of its increasingly limited ability.

The depression necessitated strict economies among farm families and stimulated a national "live-at-home" campaign to help people in agriculture survive. The NDAES grasped the value of this endeavor. As Walster explained in 1940, "every dollar that is unnecessarily spent for food, clothing, shelter, or fuel that could be produced and suitably and practically processed and manufactured on a North Dakota farm means just so much less cash available for paying debts, taxes, and for the purchase and support of everything that makes for a higher standard of living." In order to help farmers live at home, Station researchers produced a series of bulletins in 1933 and 1934 on home slaughtering, curing, and dressing of meats and poultry. Agricultural engineer Howard McColly experimented on homegrown building insulation for farms, hand-dug trench silos, and inexpensive livestock confinement systems. His colleague Bill Promersberger helped farmers adapt implements for tractors to use with horses. None of this was high-powered research, but it was what desperate North Dakota farmers needed.[22]

The drought was another problem that demanded Station attention. Animal scientists and botanists studied nontraditional forage materials, such as Russian thistle and slough hay, and agronomists paid greater attention to drought resistance in their breeding material. Shelterbelting as a method of diminishing wind erosion and conserving moisture drew the attention of several departments. McColly worked on moisture conservation through snow ridging and acquainted farmers with the techniques for constructing small dams. Entomologist J. A. Munro grappled with the resurgent grasshopper problem.[23]

While some Station researchers struggled with the economic and environmental emergencies, others chronicled the effect of those problems on the people of the state. E. A. Willson, for example, detailed the deteriorating standard of living among farm families, and Cap Miller and Willard Brown illustrated an ominous rise in farm tenancy. Sociologist Donald Hay pointed out that outmigration had been insufficient to prevent "a piling up" on North Dakota farms of young people who lacked opportunity elsewhere. The social and economic news from the Station was glum during the thirties, but it was accurate.[24]

In strange and unhappy ways the thirties provided support for many positions the Station had taken over the years. Diversified farmers were more likely to survive than their undiversified brethren, as were those attuned to advanced management techniques. Dairy specialist J. R. Dice happily noted the increase in dairy production in the state. What he failed to point out was that many desperate farmers turned to dairy cattle as a means of having something that would produce income, not because the Station's arguments had finally persuaded them. The Dust Bowl seemed to vindicate traditional Station skepticism about agriculture on the Missouri Slope. Honest to the point of being blunt, Walster did not think that even the new tillage techniques being introduced by the Soil Conservation Service would make agriculture feasible in much of western North Dakota. The drought proved to him that the future of the Slope was as a range cattle region.[25]

Being proved right did not comfort the Station much during the thirties. There were too many problems and too much pain in the state. Difficulties piled up, one upon another. Just when it seemed things could get no worse, they did. In 1935, as already mentioned, a virulent strain of rust hit the state, wiping out much of the spring wheat crop in one of the few years during the decade when there was sufficient moisture. Ceres, supposedly rust-resistant, proved especially susceptible, while Thatcher, a recent release by the Minnesota station, held up quite well. The 1935 rust outbreak ultimately benefited the Station because it led to the development of new varieties and to the important realization that there were different "races" or strains of rust. But these blessings were well disguised in 1935. Angry farmers blamed the Station for supposedly pressing them to use a variety that was not really rust resistant, and the agronomists themselves must have wondered about the quality of their work and its future.[26]

III

It must have seemed during the thirties that everything was against the Station. Depression and dust, low salaries and miniscule operating budgets, rust and grasshoppers—all of these plagues and more struck within a few years' time. Then, just when it seemed fate had turned against the NDAES in every possible way, yet another affliction appeared. Fate dealt the Station Bill Langer.

William Langer is the major figure in North Dakota political history.After graduating from the University of North Dakota Law School and Columbia University—in that order—the brilliant and dynamic Langer established an enviable reputation in private practice and the State's attorney's office in Morton County. Langer shrewdly saw the potential of the infant Nonpartisan League and succeeded in gaining election as state attorney general in 1916 under its banner. He broke with the League after World War I and helped its conservative enemies weaken it and remove it from power. This episode established for Langer what proved to be a lifelong reputation for duplicity and double-dealing, but by the late twenties he had reconciled with a number of League figures. Through liberal use of his own money he transformed the NPL into a personal political machine.[27]

Langer was elected governor with League support in 1932. Shortly after his inauguration in 1933 he moved imaginatively to solve the farm crisis, declaring a moratorium on mortgage foreclosures and an embargo on wheat shipments from the state. The effectiveness of these actions was minimal, but Langer gained much genuine admiration and gratitude in North Dakota for vigorously attacking problems when so many public officials seemed paralyzed. In 1934 Langer was removed from office by the state supreme court when a jury convicted him of illegally soliciting campaign funds from federal employees. His conviction was later overturned, and he was elected governor a second time in 1936. In 1940 he was elected to the United States Senate, where he served as an unpredictable and sometimes zany maverick until his death in 1959.

Like Louisiana's Huey Long and Minnesota's Floyd Olson, other charismatic and controversial figures who served during this time of crisis, Langer stirred strong and dramatically conflicting emotions. To his many champions he was a man of the people who was sincerely devoted to their welfare and was willing to do whatever was necessary for them. To his detractors he was a dangerous demagogue willing to trample statute and constitutional law underfoot in his single-minded drive for personal power. He was a man reasonable people could love or hate, and friends and families became bitterly and sometimes permanently estranged over him. He was not a person about whom it was easy to be neutral.

Langer's years as governor saw a degree of political interference in the College and Station rivaling that during the nineties and the teens. The governor attempted to make the NDAC and NDAES a vehicle to support his programs and a source of patronage and political funds. And they were not unique in this regard. Langer tried to gain control of every state agency that might advance his program or his political career.[28]

The first sign of increased partisan interference in College and Station affairs appeared shortly after the 1932 election when the League-dominated Republican party solicited NDAC and NDAES employees for funds to help it retire its campaign debt. This request, ignored as insulting by most people facing substantial salary reductions, was but a harbinger of things to come. When he assumed office Langer pressed employees to devote 5 percent of their salaries to purchasing subscriptions to the *Nonpartisan Leader*. President Shepperd argued that neither employees of the Board of Administration nor those who received some of their salary from the federal government could be made to contribute, but Langer refused to give up. He engineered the appointment of Steve Hagan, the son of one of his trusted lieutenants, John Hagan, as secretary of the College in 1933. The younger Hagan, a former prize fighter, persuaded many staffers to buy *Leader* subscriptions. When Chris Jensen contended that "Steve Hagan was Langer's hit man" on campus he was using the term in the literal as well as the figurative sense.[29]

Steve Hagan also had the task of uncovering or generating negative information about people on campus considered by Langer to be enemies. He was not the only person with this function. Langer also appointed Clarence McCulloch as "special examiner of College and Station affairs." McCulloch's function was essentially the same as that of Kemper Peabody, who had been appointed to a similar position by Governor Eli Shortridge forty years before—he was to find evidence of malfeasance that the governor was certain was there. McCulloch focused on Director Trowbridge, and he reported great progress in his investigation. In the first half of 1934 he collected numerous affidavits, mostly from Station laborers, that accused the director of misappropriation of crops, livestock, equipment, and labor from the Station for his personal benefit. Whether these charges were true cannot be determined. There was never any formal proceeding against Trowbridge, though it is highly probable that his agreement to become director emeritus on July 1 was made in return for Langer not leveling public charges. Whether the accusations were well grounded or not, Trowbridge, aging and suffering from heart disease, may have decided to withdraw discreetly and avoid a fight in which his honor and integrity would be questioned.[30]

The removal of Trowbridge was not sufficient to bring the NDAC and the NDAES to heel. The staff continued to refuse to endorse political positions held by Langer, and, protected by Shepperd, many resisted contributing to the governor's machine. Langer's supporters maintained a steady stream of criticism, arguing, among other things, that the College was engaging in inappropriate educational endeavors, especially by offering instruction in the liberal arts. This disingenuous criticism ignored the specific requirements of the Morrill Land-Grant College Act, but it served a political purpose. It was endlessly and cynically repeated by political enemies and educational rivals of the NDAC because it appealed to the anti-intellectual proclivities of some people in the state. For its part the Station was criticized for failing to do research beneficial to the farmers and, beginning in 1935, for misrepresenting Ceres as a rust-resistant variety of wheat.

Soon after he began his second term as governor in 1937 Langer struck at his enemies in the NDAC and the NDAES. Shepperd announced in May that he wished to retire as president and return to research work in Animal Husbandry. The Board of Administration named him president emeritus on July 29 and then peremptorily dismissed seven administrators and staff members from the institution, including county agent leader N.D. Gorman, agronomist and assistant dean of Agriculture P. J. Olson, dean of Engineering R. M. Dolve, Home Economics dean Alba Bales, dean of Men I. W. Smith, registrar A. H. Parrot, and Jean Traynor, Shepperd's secretary. Walster, too well known and respected to be fired, was relieved of the directorships of Extension and the Experiment Station. He continued to act as dean of Agriculture.

Most observers saw Langer's hand behind the actions of the board and protests were immediate and vitriolic. The Fargo Chamber of Commerce and several service clubs vociferously complained about the firings. So, too, did the NDAC Alumni and Former Students' Association. Most of the county Soil Conservation Boards, with which county agents worked closely, passed resolutions calling for reinstatement of the purged staffers. The head of the North Dakota Livestock Breeders' Association implored Langer not to harm "the hub for almost all of our agricultural and livestock development work in the state," and a Fargo minister sermonized that the action of the board was similar to the behavior demonstrated by Stalin and Hitler.[31]

Langer and his supporters accused the dismissed staffers of serving the Fargo business community rather than the state, of living the high life at taxpayers' expense, and of deflecting the institution from its agricultural purposes by broadening the curriculum. However irrational these charges might have been, they appealed politically to those driven by their resentments of educated people, public employees, or Fargo, the largest and richest city in the state. The Station was criticized for representing Ceres as rust-resistant and for suggesting methods of cultivation that came to appear harmful when the Dust Bowl hit. Langer's people also attacked the NDAES on ideological grounds. Commissioner of Agriculture and Labor John Hagan accused P. J. Olson of being subservient to the grain traders and millers. During the rust outbreak of 1916 Edwin Ladd had undertaken milling and baking studies with "shriveled" or "low-weight" wheat that had been affected by rust. Ladd concluded that the traders downgraded this wheat inordinately, costing farmers a substantial amount of money. In 1935 Ted Stoa experimented again with low grade wheat, discovering that it had good germination and milling qualities, though it was low in protein. Hagen charged that Olson prevented publication of Stoa's findings because he hoped to have the Station avoid the controversy that had followed Ladd's work and because "he was working more for the grain trade than he was the farmer of North Dakota."[32]

The accuracy of Hagan's charges cannot be judged conclusively, in part because they were more matters of opinion than objective fact. There is less question regarding how credible these charges must have seemed to many North Dakotans. Without question the relationship between the Station and the millers had become much closer and less confrontational since Ladd's time. While that relationship was changing, the Station was curtailing the sort of high-visibility, public-advocacy research for which Ladd and his colleagues had become famous. Hagan seemed to be putting two and two together, and many of his auditors understandably agreed the sum was four. However credible or genuine Hagan's concerns might have been, nobody at Fargo was willing to accept them as the real reasons for the purge. Langer had friends on campus—Bolley, Herbert Hanson, Cap Miller, and Seed Commissioner E. M. Gillig, for example—but none of them spoke up for him. Most people seemed to agree that the purpose of the dismissals was to allow Langer to seize complete control. Glenn Smith assumed "that Steve Hagan pinpointed the people on the staff who would not knuckle under and fired them." The consensus was that Langer was especially eager to get control of Extension, with its network of agents, large payroll, and control over federal funds for soil conservation. But there was probably no part of the land-grant complex, or of state government in general, that Langer did not want to control. Chris Jensen's simple conclusion that "Langer was power hungry" is compelling.[33]

The purge understandably devastated morale, adding a stinging insult to the substantial injuries the institution had already suffered. Moreover, the alumni were effectively silenced when Langer threatened to end their receipt of funds through student fees, other employees who complained too loudly were dismissed, and several faculty resigned in disgust, most notably accomplished horticulturist A. F. Yeager.

The question now was who would take the helm of what threatened to become a ghost ship. Gillig and Miller sought the presidency, and Hanson had been endorsed for it by the Farmers Union and the Farmers Holiday Association. The former was the leading farm organization in the state and the latter was a militant group favoring crop-withholding and other direct actions to raise agricultural prices. Both had some influence with Langer. The board wanted to make John C. West, president of the University of North Dakota, the boss at Fargo, too. One need not be a great partisan of West to sympathize with him in this situation. He faced a dilemma so difficult as to be virtually intractable. If he took the job he would be branded a Langer stooge and suspected of being in Fargo to gut the College for the benefit of UND. On the other hand if West refused the job, Langer might retaliate against him or his institution. While his refusal might harm UND, it was doubtful that his staying away would benefit the College. Given the salary situation and the political climate it was unlikely that a qualified outsider could be attracted, and there was precious little presidential timber in the local woods. If West refused, some abject Langer toady would probably get the job, and drift, demoralization, and deterioration would become even more pronounced. West did not want the job, and nobody at the NDAC wanted him to have it, but he agreed after some vacillation to step in temporarily. Concerned about West's political independence, USDA officials C. W. Warburton and R. W. Trullinger rushed to Fargo to be assured by the new President that he would not allow the Station and Extension to be turned into cogs in the Langer machine. West assured the friends of the College that his stay would be temporary and that he had no intention of narrowing the school's scope for the benefit of Grand Forks. He was as good as his word, though he continued to be distrusted.

Even with West there the situation was awful. The North Central Association of Colleges and Secondary Schools responded to the purge by removing the accreditation of the NDAC in 1938. That was another blow to morale, and it made West's job that much harder. West was obviously in a position where winning was not possible. He deserves credit for doing the best he could for the school and the state under terribly difficult circumstances.

Replacing Walster as Station director was Herbert C. Hanson. Hanson was a Minnesota native who received advanced degrees at the University of Nebraska. At Lincoln he studied under Frederic Clements, one of the founding fathers of American plant ecology. After a stint at the Colorado station, Hanson joined the staff at the NDAES in 1930 as a botanist. Hanson was politically progressive, as were most early ecologists. He believed in land use planning and social planning as well, and thought that scientists could and should play an active role in carrying both forward. In North Dakota these beliefs put him on the left side of the political spectrum. He was a charter member of the campus chapter of the Farmers Union and was one of Langer's admirers.

Hanson's appointment as director created an awkward situation and understandably roused suspicions. His relations with many, including Walster, were correct and formal rather than cordial. Hanson did not function as a Langer tool, however. Most of the work continued as usual, and if the governor attempted to meddle in the Station Hanson apparently deflected him. Chris Jensen's judgment that "Hanson was a crony of Langer's," but that "he did not bring NPL politics into the Station" seems to have been accurate. In retrospect Hanson, like West, appears to have been engaged in damage control, and he has never received due credit for undertaking that thankless but necessary task. In 1940 he took a leave from the NDAES to run the federal government's rural rehabilitation project in the Matanuska Valley of Alaska. He never returned to the Station.[34]

IV

The only good thing about hitting rock bottom is that there is nowhere to go but up. The situation at the College and Station slowly improved beginning in 1938. In that year Steve Hagan resigned as College secretary, and with West's blessing a separate president was chosen for the College. He was Frank Eversull, president of Huron College in South Dakota. Eversull's broad background included substantial work in agriculture. In the same year North Dakota voters brought two more substantial benefits to the NDAC. Students and friends of the College, aroused by Langer's interference in its affairs, succeeded in placing an initiative on the June primary ballot removing colleges and universities from the jurisdiction of the Board of Administration and placing them under a new Board of Higher Education. The voters approved this constitutional revision. In November the voters chose Democrat John Moses over Langer-crony John Hagan for the governorship.

Still more improvements came in 1939. A number of Langer allies were removed from office by the new regime. Among these were two Board of Administration members instrumental in carrying out the purge, Jennie Ulsrud and R. A. Kinzer. Both lost their seats on grounds of malfeasance and misconduct in office. The North Central Association agreed to reaccredit the College if the dismissed staff members were offered reinstatement. The new board complied, and the North Central Association fulfilled its pledge. P. J. Olson decided not to come back, but Walster was returned to directorship of the Station. Langer's being out of the governorship made a big difference. He held public office again, but his position minimized the harm he could do to the Station. In 1940 the North Dakota voters sent him to the United States Senate. The scientists at the NDAES considered that the nation's loss was their gain.[35]

The Station could not simply wait passively for time to wound all heels. Long-neglected fences had to be mended. During the twenties Trowbridge and his associates had ended Station activities that had a clear and direct impact on North Dakotans. They had been somewhat careless about maintaining a favorable public image of the Station. Perhaps they thought the public would understand automatically how NDAES research was benefiting North Dakota. It did not. Under Trowbridge the Station had painstakingly cultivated agribusiness and conservative farm groups such as the Northwest Farm Managers Association, but it was inattentive to organizations that continued to express economic and political protest. Perhaps the NDAES thought conservative groups were the only legitimate spokesmen for agriculture. They were not. Whether the Station's failings in public relations derived from a conscious decision or simply from a lack of political astuteness, it paid for them in the thirties. Its apparent conservatism led to its condemnation by the agrarian left, and it discovered that the favorable public image it had once had was largely gone. It was easy to blame Langer, but he expressed the feelings of many North Dakotans. As Stanley Moore, former President of the North Dakota Farmers Union, notes, "there were elements in the NPL that were anti-intellectual—that's why they opposed the College . . . . On the other hand, there was a certain amount of arrogance at the college."[36]

Herbert Hanson tackled the Station's related political and image problems during his short tenure. He launched the *BiMonthly Bulletin* (now *North Dakota Farm Research*), a slick, popular publication designed to acquaint North Dakotans, in terms average people could understand, with Station research and how it benefited them. He built bridges to Langer and the Governor's supporters. He worked to develop a positive relationship with the activist Farmers Union, of which he was a member. And he encouraged critics of the NDAES to suggest how it could better serve them. It was not coincidental that milling and baking studies reminiscent of those of Ladd appeared again when Hanson was director.[37]

Hanson's solicitude for the Station's critical constituents did not prevent him from honestly apprising them of their mistreatment of the institution. He noted that state appropriations for the NDAES in the 1937-38 biennium amounted to only fifty-one cents per year for each of North Dakota's farms, "or one-ninth as much as the average amount of other states set aside for research in agriculture." If this penury continued, Hanson concluded, North Dakota would lose its markets to other, more competitive states.[38]

Hanson's initiatives were continued by Walster after he returned to the directorship. In pursuit of political support Walster assiduously cultivated groups as diverse as the Greater North Dakota Association and the Farmers Union. On behalf of the latter he went so far as to request that the NDFU receive confidential information on the USDA budget. Trusting and respecting him, Farmers Union leaders periodically discussed offering Walster a major position with the organization, such as that of director of the Grain Terminal Association.[39]

However necessary these contacts were, they inevitably carried the danger of involving the Station in inappropriate activities. Worst's difficulties with the Equity Society and its enemies illustrated the potential pitfalls of working with groups having controversial political agendas. Walster was careful to avoid these. His relations with the Farmers Union were cordial, but he refused that group's request for a public endorsement for a graduated land tax that would protect small farmers by penalizing large ones. He was no more willing to participate in politically sensitive activities at the other end of the agrarian political spectrum. In 1942 he refused to participate in an organizational meeting for a new state Farm Bureau. "Once your organization is set up and functioning, it will be clearly within our duty to cooperate with your organization," he wrote, "and let me assure you that it is the policy of the Station to cooperate with all organizations interested in the promotion of agriculture." Working freely with all interested groups without compromising Station integrity and responding to all without favoring any were challenging tasks at the NDAES, as they had always been. Walster carried them out as successfully as any director ever had.[40]

The position of the Station improved because of valuable scientific work as well as politicking and publicizing. Agronomy made especially dramatic contributions, quickly responding to the rust threat by introducing two new spring wheat varieties, Rival and Pilot, in 1939. "Moderately resistant to both stem and leaf rust [and] capable of good yields," the new varieties rapidly gained favor with farmers. Producers also seemed to be impressed with and appreciative of Agronomy's work on their behalf. During the forties that appreciation began to bear fruit in increased support.[41]

The Station's improved political position was reflected in rising appropriations. The problem was that appropriations had been so desperately low that even dramatic improvements did not have a major positive effect on the work. One legislative priority was the reopening of the substations, all but one of which had been effectively shut down during the thirties. The main station had been ambivalent about these entities ever since Worst's day. Unless they were adequately funded, which they seldom were, they diminished the efficiency of the research system rather than supplementing it. The USDA suggested that the state had too many and that some should be shut down, but the Washington bureaucrats ignored the realities of political life in North Dakota. "I appreciate the forcefulness of the argument which you set up with respect to concentrating our facilities at the Fargo Station," Walster wearily conceded to a USDA official, "but if we fail to do something of special character, such as irrigation work for people in the Northwestern part of the state, that is the very place we are going to lose the necessary legislative support which we must have in order to secure the necessary funds for the operation of the central Station."[42]

Then there was the question of projects to be undertaken. The Station had been forced to abandon many lines of work during the thirties and it was obviously eager to reopen some of these when funding permitted. Answering a memorandum from Chief James T. Jardine of the Office of Experiment Stations in September of 1940, Walster listed durum quality, barley improvement, cost-of-production studies, tenancy and rental contract studies, farm management investigation, weed control, poultry bacteria, soil and water conservation, livestock feeding and management, sewage disposal, and botulism in wild fowl as areas the NDAES would work on if it had the money. Studies in several of these fields had been terminated owing to financial exigency during the depression.[43]

Walster's main priority, of course, was improving salaries, because good work could not be done without good people and good people demanded good salaries. Walster ruefully noted in May of 1942 that the NDAES salary scale was the lowest in the United States and that, as a result, "every other state Experiment Station and every federal agency looks upon this as a recruiting ground for their own staffs." Walster admitted that that situation forced him to provide disproportionate rewards to well-trained researchers under forty years of age, "who may be expected to be in their most productive years. In order to do this, the director does not expect to ask for salary advances of the older members of the staff." One result of this necessary but unjust policy was that people who had contributed to the state for many years, such as H. L. Bolley, L. R. Waldron, O. A. Stevens, and Ted Stoa, were grossly underpaid. The financial situation in the state and at the Station thus resulted in a great cruelty and injustice to the very people who had sacrificed so long to build both.[44]

To say that things improved at the NDAES after 1937 is to speak only in relative terms. The financial situation remained serious, though no longer desperate, and Walster longed merely to get back to the comparatively flush times of the late twenties. In retrospect, though, it is clear that the Station was entering a period of political popularity and financial strength unknown in its previous history, a period that would last for half a century.

Thomas P. Cooper, about 1915.

Perry Fox Trowbridge, 1934.

Harlow Leslie Walster in his spartan office, 1926.

Walster (far left), and John Lee Coulter (far right), entertain visitors on a Station alfalfa plot in 1927. They are, from left to right, A.G. McCall of the USDA, an unidentified Soviet soil scientist, and Sir Henry Russell, director of the Rothamsted Experiment Station in Great Britain.

O.A. Stevens, 1932.

Cereal technologist C.E. Mangels, 1934.

Testing wheat for protein, 1928.

Plant pathologist W.E. Brentzel, 1934.

Bacteriologist Casper Nelson, 1934.

H.L. Bolley showing visitors around the plots in his characteristically energetic manner, 1924.

Lawrence Root Waldron, about 1935.

Patents Kota and Marquis flank illustrious offspring Ceres.

Agronomist Ted Stoa in a flax field, 1934.

Intrepid entomologist J.A. Munro examining bees, 1934.

Horticulturalist A.F. Yeager, 1934.

H.F. McColly of Agricultural Engineering, 1933.

Why so many got away: One of Entomology's state-of-the-art insect sprayers in 1934.

Barberry eradication squad in front of Science (now Minard) Hall in 1921.

This 1926 shop scene shows that the agricultural engineers were keeping up with the mechanical revolution.

Dairy specialist J.R. Dice touting better practices in Leonard, 1928.

Northwest Farm Managers Association tour in front of recently completed Morrill Hall, 1926.

Esther Latzke's cooking lab, 1929.

Sign of the times at the NDAES during the thirties.

Herbert C. Hanson, 1936.

College Secretary and Langer lieutenant Steve Hagen, about 1935.

William Langer in the early forties.

7

PLANNING FOR THE FUTURE AND THE PAST

The economy of the State of North Dakota is in a state of flux. The impact of the rapid development of hydroelectric power, the discovery of petroleum, the improvement of highways and the greater use of truck transportation, the almost complete mechanization of power units on all farms and the application of those power units to every type of farm machine, and the widespread use of a great variety of agricultural chemicals for protecting crops and animals, for fertilizing the soil, and for killing weeds, all sums up to an enormous rapidity of change.

H.L. Walster, 1952

To most of the farming population, these last ten years have been only a chance to regain lost ground, and a chance to prepare for the uncertain perils of drought or depression which might recur. Nature is a rough playmate in the Northern Great Plains, and the rewards in the market place are uncertain.

Baldur H. Kristjanson and
C.J. Heltemes, 1950

The 1940s and 1950s were a golden age for publicly sponsored agricultural research in the United States. During the thirties many students of public policy had wondered whether, in light of oppressive surpluses, there was any justification for scientific developments that seemed to increase production. Spokesmen for government research, capably led by Secretary of Agriculture Henry A. Wallace, defended the system intellectually, but it took no less a figure than Adolf Hitler to prove its practical worth. When Hitler and his allies forced war upon the world, the surpluses that had so stubbornly resisted the puny depression-era assaults of man and nature melted away. Overnight, American abundance became a source of comfort rather than consternation and accomplished scientists and productive farmers were transformed from goats to heroes. Americans were reminded again that, if having too much is a problem, it is far from the most serious problem a people might have.

Many people in the research system during the early forties remembered the previous world war. It had presented few positive results for either farmers or agricultural scientists. This one turned out differently. World War II vastly increased the prestige of science in the United States. Science seemed to point the way toward human domination of nature and American domination of the world. Agricultural science promised a cornucopia beyond the grasp of all but the most fertile imaginations. And young men and women shared the vision, crowding into graduate schools by the thousands in an attempt to be part of the great vanguard of science.

Agricultural science seemed immune to criticism in those happy years. When surpluses returned, science was seen as the solution more than the problem. Scientists could turn food and fiber into industrial products, and social scientists could devise more efficient marketing mechanisms. What was left could be painlessly given by a rich nation to poor peoples; perhaps in gratitude they would follow us rather than the Soviets, our cold war adversaries. Even the rural depopulation that inevitably accompanied dramatic increases in farm productivity failed to reflect negatively on the research system. There were plenty of jobs in the booming cities, went the argument, and people were better off living there in prosperity than on farms in poverty.

North Dakota participated to some extent in the national trend. High demand and ideal growing conditions in the forties gave the state's farmers a level of prosperity they had rarely enjoyed. Electricity transformed their lives, wiping away the major distinction between rural and urban lifestyles.Their children joined the service and went to high school and college in record numbers, eroding the parochialism of local areas. Adults broadened their horizons by traveling, both to towns and cities in the area on improved roads and to other states and regions. North Dakota farmers mechanized, expanded their operations, improved their agricultural and business practices, and participated in the chemical revolution, becoming dramatically more productive in the process. As Director Walster observed in his annual report in 1952, the state and its agriculture were undergoing "an enormous rapidity of change."[1]

But nothing really changes completely, and North Dakotans and the people in the North Dakota Agricultural Experiment Station who served them recognized that reality. North Dakota remained an isolated state with a challenging environment. There was no reason to think that the old problems would not return. As agricultural economists Baldur Kristjanson and C. J. Heltemes reminded readers of a June, 1950, bulletin, good times provided "a chance to prepare for the uncertain perils of drought and depression which might recur. Nature is a rough playmate in the Northern Great Plains."[2]

So, like old soldiers, the scientists at the NDAES prepared to fight the last war. The enemies were fluctuating wheat prices, drought, and rust. The weapons they grasped included diversification, irrigation, and an active breeding program. Like old soldiers, they sometimes did not see that their enemies had been neutralized, that government programs, for instance, had provided unprecedented stability for wheat prices. But their prudence and conservatism stood them in good stead, because some of the old enemies were still around and it was good to be ready for them.

I

For those in the public research system the most remarkable aspect of World War II was the dramatic turnabout it brought in the popular American attitude toward agricultural science. During the thirties abundance was more bane than benefit, and those who had helped create it did not enjoy a favorable public image. The onset of war led Americans to cherish their surpluses and to celebrate those who helped produce them. "A short while ago some people felt it [agricultural science] had done its work too well," noted USDA research administrator E. C. Auchter in 1942, but "to-day there is no doubt in any one's mind about the need for science in agriculture. Vast armies have to be fed, clothed, . . . and kept healthy. Our allies need enormous quantities of agricultural products. Our industrial workers must be well nourished to produce efficiently."[3]

Agricultural scientists and farmers were able to respond quickly when the nation called. Agriculture was at last able "to use to the full the vast reservoir of scientific knowledge and technological know-how which had been accumulated during the nineteen thirties, but which was then held back from use because of drought and depression." Probably the most important development was that of hybrid corn, which was exploited to its full potential during the war after being developed during the twenties and introduced during the thirties. When "it became necessary to pit our total strength against a coalition of ruthless, lawless aggressor nations in a life-and-death struggle," one breathless commentator remembered, "American corn was projected into unprecedented importance around the world." Of greater long-range significance were new agricultural chemicals. The insecticide DDT, as well as 2,4-D, a remarkably effective herbicide for broadleaf weeds, were introduced during the war. Anhydrous ammonia, which revolutionized nitrogen fertilization, came shortly afterward.[4]

Agricultural science benefited not only because it was contributing to greater production but also because it was science. As never before Americans came to appreciate the possibilities of science and technology during World War II. "Science is our miracle worker," noted Cornell University president Edmund Day in 1941. "It is to science that we look for relief from practically all our ills." During the war science was celebrated for its many contributions to our victory, the most dramatic of which was the atomic bomb—so awesome and terrible to a later generation but awesome and wonderful to many Americans in 1945. The war illustrated the potentials of science, and Americans afterward expected myriad peacetime miracles.[5]

Expectations for miracles in agricultural science were extravagant. Prognosticators forecast revolutions in plant and animal genetics, soil and water conservation, agricultural chemistry and farm machinery. Popular authors assured readers that photosynthesis would be controlled, atomic energy would be applied to farm operations, and insect pests would be obliterated. Nobody put agricultural science's light under a bushel any longer.[6]

The benefits that accrued to agricultural science after World War II were not confined to the realm of image. As Norwood Kerr notes, "the esteem accorded agricultural science directly benefitted the state experiment stations in two major ways: a whole new generation of young scientists was attracted to public research and public financial support increased dramatically." The component parts of the research system thus received a needed infusion of talented and idealistic young scientists as well as impressive increases in funding.[7]

While agricultural scientists were understandably proud of their contribution to society and optimistic about the future of their enterprise, the dramatic alterations in agriculture to which they were contributing were mixed blessings. More advanced agriculture demanded more capital and greater business and management skills of farmers. Greater productivity narrowed profit margins and placed a premium on economies of scale, forcing many struggling producers out. Machinery cost hundreds of thousands of laborers and sharecroppers their jobs, and chemicals posed new economic and health challenges to farmers.

Those who raised questions about the direction of change or the responsibility of the research system for it did not receive a very sympathetic hearing. People like Ralph Borsodi and Whitney Griswold who hoped to preserve farming more as a way of life and less as a business were commonly dismissed as hopeless romantics. Rural sociologists and agricultural economists who complained about the effects of industrial agriculture on farm workers, small producers, or rural communities were generally silenced or forced out of the system, to be replaced by social scientists asking safer questions about why some farmers innovated while others did not. The unwillingness of the system to tolerate dissent was unfortunate and ultimately harmful, leaving it ill prepared to respond effectively to more scathing criticisms in the sixties and seventies.[8]

In the postwar years the champions of the system saw only positive aspects in agricultural change. There were fewer farmers, they conceded, but those who remained were more intelligent and were better businesspeople and managers than farmers ever had been. The rural community was shrinking, but the people there lived better lives than ever before. Costs were high, and surpluses had returned, but science held the solution to problems it had helped create. If the world's hungry were fed, if marketing were improved, if farmers became better managers, and if agricultural products could be utilized for industrial purposes, all of the problems of agriculture would melt away. For the agricultural research system as a whole the forties and fifties was an ebullient time, and there was no inclination to dispel the happy mood.[9]

II

In retrospect, World War II can be seen as a great turning point for North Dakota agriculture. Rising prices and the end of drought conditions conjoined to bring unprecedented prosperity to North Dakota farmers. Depression-era federal price-support programs, which increasingly assumed the appearance of permanence, and such other government programs as crop insurance and low interest loans, foretold a much more stable agricultural environment. Improvements in farm technology and the dawning chemical revolution promised to increase the individual's productivity dramatically. Electrification and improvements in transportation presaged a rural life with all the material comfort and stimulation of urban life. From our perspective, North Dakota farmers were poised to move briskly into the modern age. They did not often see it that way.[10]

Rural North Dakotans tended to view the years of the early forties as an aberration. They had lived through such an economic and environmental hell that they could not see the future outside the context of the past. The rains had banished the drought, but it could easily return. Prices were high, but that had been true before and they had fallen. Government programs provided some security, but they could easily be wiped out by Congress. Machinery, chemicals, roads, and electricity were all very nice, but all had to be paid for.

The actions of North Dakotans illustrated how they continued to be gripped by the psychology born of drought and depression. Good weather and high prices did not lead them astray. Farmers with mortgages paid them off. Farmers without mortgages put money in the bank, preparing for the inevitable un-rainy day. Purchases of land and machinery were conservative and carefully calculated. In 1932 the legislature had outlawed corporate farming and had decreed that existing corporations must divest themselves of farmland within ten years. When the deadline came in 1942 banks and insurance companies threw thousands of attractive farms on the market. But buyers were prudent, even in the face of this rare opportunity, and were fearful of overextending themselves by buying too much land or paying too much for it. As a consequence, much land went unsold or was acquired for a song. During the twenties and thirties farmers had struggled to keep their heads above the surface in a sea of debt. Many had not succeeded, and most of those who had were reluctant to plunge in again. They watched their nickles and dimes, living as self-sufficiently as they could and limiting commitments to better schools and roads and state services, all of which would cost money.

Rural North Dakota during World War II was like a patient recovering from

explain rather no vote of 1989

a serious illness. Not only was it intensely afraid of a recurrence, it also continued to suffer aftereffects. Many of these were chronicled by a remarkable series of studies undertaken by agricultural economists and rural sociologists in the NDAES and in the USDA's Bureau of Agricultural Economics. In June of 1940, for example, Station sociologists highlighted the problem of rural overpopulation they had perceived initially in the mid-thirties. Donald Hay, James Greenlaw, and Lawrence Boyle noted a glut of young adults in rural areas due to the constriction of urban employment opportunities. As these researchers noted glumly, "the total numbers and proportion of youth in the population of North Dakota have reached an all-time high in a period when there is no longer any free land, when depression and drouth have greatly reduced the agricultural income of the State, and when there is a scarcity of opportunities elsewhere." Hitler and Tojo rapidly took care of the population surplus, and by July of 1942 Greenlaw and two associates could report that "the requirements of the Selective Service Act and voluntary enlistments have accounted for some of the increased loss of farm population and movement to war industries has accounted for additional losses." That movement was good for the migrants and for the country, but it increased the likelihood of "a smaller, and probably older, farm population," and "a reduction in the numbers of businesses and of social institutions such as schools, clubs, churches, and . . . government agencies." Many of these were already suffering because of years of rural hardship and depression-era tax moratoria.[11]

Farmers were too preoccupied with their own recovery to worry much about support for their local institutions. One investigator from the Bureau of Agricultural Economics noted that "crops are good, agricultural prices high, bank deposits large, credit plentiful, [and] interest rates low," but still farmers were mainly interested in "paying debts, some of which are the result of their land purchasing experience of World War I." They were apparently reluctant even to spend much to improve their own or their families' comforts. In 1945 fewer than 10 percent of North Dakota farm homes enjoyed running water, and barely one-third had telephones. As late as 1949 only 36 percent of farms had central station electric power, ranking the state the lowest in the country.[12]

Like the farmers it served, the Station continued to be shaped by the experience of the thirties. Walster and his staff continued to stress prudent diversification and to urge the return of much of western North Dakota to grass. L. R. Waldron and the agronomists continued to search for high-yielding, high-quality wheats that were drought- and rust-resistant. The Station continued to be handicapped by a lack of money and staff. Appropriations rose modestly but remained woefully inadequate given the tasks at hand. Only reluctantly and slowly was the legislature brought around to the realization that it must stop reappropriating federal funds as a means of avoiding its own responsibilities for the maintenance of agricultural research. Financial difficulties contributed to a severe shortage of scientists during the war. As Walster lamented to Governor John Moses in February of 1943, "our agricultural manpower, and our potential agricultural leadership continues to be fed into the Army, Navy, and so forth at a rapid rate."[13]

The shortage of money and personnel and the press of war work had a decidedly negative effect on the Station. It prevented researchers from undertaking new lines of investigation or even reopening old ones closed during the depression. Indeed, the scope of Station work had to be constricted further. In order to devote funds to production research and to the upgrading of some salaries Walster was forced to terminate James Greenlaw and his associates in Rural Sociology. As a hard scientist with minimal respect for the social sciences, Walster no doubt felt he could most easily afford to do without Rural Sociology. But no administrator likes to lose researchers, especially when they are talented practitioners in their disciplines.[14]

There were also some very positive developments at the NDAES during World War II. The most important of these was the introduction of Mida wheat. The result of nearly thirty years of wheat breeding at the Station by L. R. Waldron, Ted Stoa, Glenn Smith, and their associates, it was the best variety developed at the Station up to that time, combining rust-resistance and good yields with satisfactory milling and baking qualities. Mida quickly became a favorite with North Dakota farmers, putting millions of extra dollars into their pockets. Mida was crucial in improving the Station's position with the legislature. "Prospects for liberal action on the part of the legislature for the benefit of the Experiment Station are very good, indeed," Walster wrote James T. Jardine of the USDA's Office of Experiment Stations early in 1945. "Our plant breeding program has really registered in this State in a remarkable way. There is an equally insistent demand now that we do as well with some of the other crops as we have done with wheat." The breeding program and the patient attention of dedicated NDAES scientists to it had finally borne fruit. In the forty years after World War II it became the source of most of the Station's fame, many of its greatest accomplishments, and much of its legislative and popular support.[15]

The value of the wheat-breeding program at the NDAES was made abundantly clear by the outbreak of 15-B stem rust, beginning in 1950. Rust was an old enemy of North Dakota farmers, and it had an uncanny way of appearing at the worst possible times. A rust outbreak in the mid-teens, for example, prevented North Dakota farmers from taking full advantage of high prevailing prices caused by the outbreak of World War I. Another devastating attack in 1935 ruined one of the few promising crop years in that dismal decade.

The NDAES had done all that it could to combat this parasitic menace. L. R. Waldron had started breeding for rust resistance in 1916. Observing that barberry bushes provided a trysting place for amorous rust spores, H. L. Bolley had prodded the USDA into eradicating that common shrub. The 1940s seemed to show the success of these and other endeavors as North Dakota enjoyed a decade of high yields and sharply diminished losses to rust.

But, as Dickinson superintendent Tom Conlon wryly cautions, "Mother

Nature keeps you honest. She is more proficient at developing plagues than we are at developing solutions." And so it was with rust. The problem was that there were over 200 varieties, or "races," of rust. As one expert noted in 1949, "a wheat variety may be immune from some races, moderately resistant to others, and completely susceptible to still others." Compounding the problem was the fact that "new races are produced continually by a process of hybridization between existing races." When rust struck it usually did so very quickly, with little warning, and with devastating effect. Leeds farmer Al Kenner recalls the rapidity with which 15B struck his durum crop: "I'll always remember July 4, 1953 . . . . We had found a few spots a few days before . . . . We decided that day the crop was going to be rusted out. I tried to salvage a little bit of seed from one field of foundation Mindum durum... and from a fifty acre field we got 150 bushels of 43-pound durum. It was just chicken feed."[16]

Race 15B was especially frightening because, as Glenn Smith noted, it was "a race so virulent that it attacks any wheat now grown commercially." Durum varieties, resistant to most rusts, were especially vulnerable to 15B. Fortunately, the NDAES was not without resources in combatting the 15B threat. Because it had carried on extensive wheat breeding for thirty-five years it had a wealth of genetic resources on which to draw. Even varieties that had not proved commercially feasible could be used as breeding material if they showed resistance to 15B. So many varieties were regularly planted on plots or in greenhouses by Station agronomists that already in 1950 some relatively resistant varieties were identified. Most promising was a Palestinian durum variety. Agronomist Ruben Heermann began crossing it with local durums that were high in quality but susceptible to rust.[17]

The Station moved quickly to increase the crosses that carried the resistant characteristics of the Palestinian parent. Ted Stoa conceived the idea of taking two of the resistant lines to Brawley, California, for winter increase. This winter increase program allowed the Station to double generations of resistant wheat and vastly increase available seed. As seed stocks specialist David Ebeltoft has noted, "that put us way ahead of the other states" struggling with the rust emergency. The Station was well circumstanced for summer increase as well. For many years the NDAES had enjoyed a close relationship with the North Dakota Crop Improvement Association, an organization of growers of certified seed. In order to provide the means for more rapid increase of new varieties, the Crop Improvement Association contributed $120,000 to be used to acquire land for an Agronomy Seed Farm. A section near Casselton was purchased in 1950, and the Agronomy Seed Farm went into operation. Together the Agronomy Seed Farm and the branch stations made the rapid increase of varieties resistant to 15B possible.[18]

Throughout their long and difficult battle with 15B, Station agronomists enjoyed a heartening degree of public support. After 1935 John Hagan and others had berated the NDAES because its favored variety of hard red spring wheat, Ceres, had proved vulnerable to the rust outbreak of that year. Those criticisms struck a responsive chord in some segments of the public. In 1950 the public attitude reflected much greater sophistication and understanding. "It was interesting to me that the farmers did not criticize the Station for not being ahead of the rust," Glenn Smith, who remembers the 1935 outbreak, has remarked. "That was the amazing thing to me—they were really very supportive." Smith attributes this altered attitude to the work of Ted Stoa and others, who patiently educated farmers regarding the true nature of wheat rust and the difficulties of developing resistant varieties that were also satisfactory in yield and quality. But changing times were important, too. As the desperate thirties faded into memory their bitter and sharp-edged politics also faded. For its part, the Station under Walster had carefully crafted a favorable public image and a good relationship with politicians, farm organizations, and agribusiness, through both its accomplishments and its increased willingness to publicize those accomplishments. Because the NDAES learned the lesson of the thirties it was not compelled to repeat some of the painful experiences of those years.[19]

The improved public position of the Station paid dividends to it and to the state it served during the 15B crisis. The NDAES had lots of highly placed and organized friends and few enemies. J. W. Haw of the Great Northern's Development Department, whose interest in North Dakota agriculture stretched back to the days when he was a Better Farming Association agent, pressed Governor Fred Aandahl for emergency funds for the Station's winter increase program. Senator Milton Young, diligently cultivated by Walster,demanded extra money from the USDA. The Rust Prevention Association, composed of companies in the grain, milling, and baking businesses, contributed funds to the Station and lobbied for it in Bismarck. And North Dakota Farmers Union president Glenn Talbott promised "help . . . to resolve the problems in the years ahead in regard to needed appropriations." The public response to the 15B crisis indicated the degree to which the Station had worn down, outlived, neutralized, or converted its enemies.[20]

Race 15B was not defeated quickly or easily, despite the impressive response of the NDAES. After relatively rust-free seasons in 1951 and 1952, 15B returned with a vengeance in the next two years. In 1955 the Station took the unprecedented step of simultaneously releasing four resistant durum varieties. Langdon, Yuma, Ramsey, and Towner. Lee, a hard red spring wheat, was released cooperatively with the Minnesota station. Selkirk, a resistant hard red spring wheat from Manitoba, was released in 1956 after testing and increase by the NDAES. By that time sufficient quantities of resistant, agronomically suitable, adequately yielding varieties were being grown that 15B faded away as a problem.

The NDAES had played a major role in developing and increasing the rust-resistant varieties, and it received much credit for its accomplishment. To those involved in this effort, and to those who remember it, the battle against 15B marked a great turning point for the Station. Wayne Owens, for example, who was Ramsey County agent in the fifties, remembers that the "new durum varieties that were rust resistant . . . created a great deal of awareness among farmers of the value of the Station." Fessenden farmer and veteran state senator Bryce Streibel says simply that the "Station has made an economic contribution to the state without peer. When rust took out the durum, our economy would have been devastated were it not for our research."[21]

In addition to portraying the Station in a very favorable light, the 15B episode slightly altered the public perception of its rôle. The NDAES came to be seen preeminently as a plant-breeding institution. Today nine of ten knowledgeable observers will mention new varieties first when asked about the outstanding contributions of agricultural research in North Dakota. This image has both its positive and negative aspects. It gives the Station a very visible means of illustrating its service to the state. The Station has devoted substantial resources to plant breeding and the investment has paid off. A steady stream of new varieties of wheat, barley, oats, potatoes, sunflowers, corn, soybeans, and other crops has flowed out of greenhouses and into North Dakota farmers' fields over the past generation. These tangible and regular introductions have been effective means of maintaining and increasing legislative support. On the other hand, the visibility of plant breeding has partially obscured other, less spectacular contributions made by the Station in such areas as soils, weed and insect control, and animal science, among others. It has also put the Station in the uncomfortable position of being a potential competitor with private enterprise. Finally, it has placed pressure on the Station continually to introduce improved varieties. People who remembered yesterday's miracle were impatient to see today's. Of course, the alternative to the situation arising from the 15B fight would be worse. Had the Station failed to respond and serve the state then it would feel no pressure now. People expect accomplishments only from the successful, and that was what the NDAES showed itself to be.

The appearance of race 15B reminded the Station and the state of the fragility and unpredictability of an economy based on wheat monoculture. As it had since 1890, the NDAES continued after the war to search for the means of reducing risk in the state's agriculture. Agricultural economist Rainer Schickele grappled with the economic and policy aspects of risk. Within a few years after he came to the NDAES in 1947 Schickele had built a department of bright and active young scholars. Under his leadership Agricultural Economics gained some national recognition and also won some respect from Station researchers skeptical about the social sciences.

Schickele saw wheat monoculture as the basic economic problem of North Dakota. Wheat was a risky source of income because "in extensive grain farming, the real cost per crop acre remains nearly the same regardless of the wide variations in production." In North Dakota rust, hail, drought, and other natural factors contributed to production variability, and the charges of middlemen and wide price fluctuations added to the unpredictability of profits. Schickele believed it was necessary to "find ways, through a combination of individual practices, cooperative organization and public policies, for a competent and honest farmer *to ride out periods of drought and poor yields without starving his family, robbing his soil and losing his farm.*" He also believed it was necessary "*to provide an adequate minimum of economic security over the years for any farm family willing to do its share*, in face of the extreme crop and price hazards to which the region is exposed. That is a new frontier for Plains farmers to conquer." In the years that followed, Schickele and his colleagues closely studied such areas as cooperation, crop insurance, variable rental and mortgage plans, and federal price support policies in their search for enhanced stability for North Dakota farmers.[22]

While the agricultural economists studied the policy and management components of income stabilization, others at the Station continued to pursue that goal through diversification. Livestock remained the favored road to diversification at the NDAES. Cattle, hogs, dairy products, and poultry provided steady income, efficiently consumed forage and low-grade grain, and helped maintain soil fertility. The Station championed livestock especially in the west, where low average rainfall made undiversified grain farming particularly precarious.[23]

Crops other than wheat also promised a more diversified agriculture. Led by the brilliant USDA scientist Harold Flor, the Station continued to be a major center of flax research. Over the years the NDAES had been involved in the development of several varieties of high-yielding, wilt- and rust-resistant flax. Ironically, it was the development of petroleum-based paints, varnishes, and floor coverings after 1960, not the ravages of rust and wilt, that led to the decline of flax.

Barley was a different story. It is an attractive crop for many North Dakota farmers because it is similar to wheat and thus requires no special equipment or skills. Another desirable feature of barley is that as an ingredient in brewing and distilling and as a nutritious animal feed, it enjoys two major markets. North Dakota farmers started to grow it in quantity in the teens. At the same time the NDAES initiated efforts to improve barley quality. In these early years farmers paid little attention to maltsters' specifications, and the onset of nationwide Prohibition in 1919 dramatically constricted the most lucrative domestic market for barley anyway. The end of Prohibition, severe declines in yield in other producing states due to disease, and greater attentiveness by the state's farmers to brewing specifications made North Dakota a center of barley production in the early forties. Within a decade maltsters were showing a preference for North Dakota barley, and the American Malting Barley Association was contributing funds to support Station research on the crop.

Potatoes were another crop that appeared about the time of World War I. By the late forties, with the support and encouragement of the Red River Valley Potato Growers, the NDAES was conducting a vigorous potato research program focusing on disease and insect control, fertilization, and mechanization. A breeding program that eventually proved remarkably productive was also under way.

Sugarbeets had been investigated by Edwin Ladd before the turn of the century. Ladd proved that beets could be grown in the Red River Valley, but it was not economically feasible to do so until a local factory was constructed to extract the sugar from them. That occurred in 1926 when the Minnesota Sugar Company built a plant at East Grand Forks. The postwar years saw an upsurge in research on beets, concentrated especially on mechanization and insect problems. In the mid-fifties the Red River Valley Sugarbeet Growers Association began providing support for Station research on their lucrative crop. Research in new crops such as soybeans and sunflowers also began soon after the end of the war.

A broader crop mix promised to be an important factor in stabilizing the incomes of North Dakota farmers. More stable incomes would also result from altering or controlling the environmental conditions under which crops were grown. Nobody at the Station after World War II who had been there before could forget the devastating drought of the depression years. These unhappy memories were reflected in a strong research effort involving water. Walster showed a particular fascination with the climate of the Great Plains and its possible modification. He followed the tree ring studies of Bismarck nurseryman and seedsman George Will with great interest, hoping that they held the key to predicting long-term weather patterns. He was also interested in weather modification—at best a pseudoscience only a few years before—which was studied at the Station in a modest way after World War II.[24]

More promising in the short term was irrigation. Walster had been intimately involved in the formulation of the Pick-Sloane Plan for water development in the Missouri River Basin. In essence, this plan called for the erection of dams on the Missouri in the upstream states of Montana and North and South Dakota. These impoundments would provide flood control beneficial to the downstream states as well as electricity and water for irrigation for the upstream states.

Many in the state saw great potential benefits in this development, even after it became apparent that most soils were inappropriate, dashing the hopes of irrigation enthusiasts that Missouri River water would solve the problems of arid western North Dakota. The idea of irrigation survived in a plan to divert water from behind the Garrison Dam northwest of Bismarck to the eastern part of the state through a complicated system of impoundments and canals. The Garrison Diversion Project has not been completed and probably will not be, much to the distress of many North Dakotans bitter about what they see as bad faith on the part of the federal government.

Nobody foresaw this disappointing outcome in the early postwar period, however, and the Station busily prepared for irrigation, studying it in a variety of settings. In 1957, with the strong support of the Garrison Diversion Conservancy District, the NDAES got the legislature to appropriate funds for a special branch station at Carrington to undertake irrigation research. As always, the Station believed the state was best served by honesty, and it never portrayed irrigation as the magic bullet for which many North Dakotans took it. Researchers were careful to point out that it was expensive and would not work on all farms or soil types, but they also saw it as a means of mitigating climatic problems and stabilizing farm income.[25]

This enduring preoccupation with stabilizing farm income—along with special federal funds that came with the passage of the Research and Marketing Act of 1946—led the Station into expanded marketing research after World War II. North Dakota farmers and the agencies and organizations that served them had not been much concerned historically with marketing problems. Walster thought attention to the issue was long overdue, noting in 1950 that "the agricultural production of the state must continue to seek its chief market beyond its borders. In that market it must compete with the nation and the world."[26]

Some of the news that Station marketing researchers had for North Dakota farmers was not very encouraging. They pointed out that distances prevented North Dakotans from effectively marketing fluid milk and that sparse cow population and infrequent deliveries kept butter quality low. Egg quality was also low, for some of the same reasons. The Station sought to rectify the dairy problem by instituting a program to improve butter quality. That effort had some marginal successes, but the intractable difficulties of dairy and egg marketing, increasingly apparent in these years, led the NDAES to de-emphasize research in both areas. In other areas marketing research could point producers in a more positive direction. For example, when Leroy Schaffner showed in 1948 that consumers preferred "more carefully graded and packed" Western tablestock potatoes over those marketed from the Red River Valley, he gave local producers a useful and potentially profitable piece of marketing information.[27]

III

Many of the emphases of the Station in the postwar years paralleled those of earlier times. Income stabilization, diversification, and wheat breeding were all areas of enduring interest at the NDAES. Such continuity was both understandable and justifiable. Research programs cannot be turned on and off, and scientific breakthroughs cannot be produced on demand. A breeding program has to proceed for decades—even generations—if it is going to be successful. There are times when it seems unnecessary, but it must be continued if it is to be fruitful when times change. The search for a rust-resistant wheat seemed irrelevant during the forties, but the 15B crisis showed its value. The NDAES looked backward at an agricultural system based on the production of a crop vulnerable to economic and natural threats of great severity. The Station prepared for the past, and 15B showed that was not necessarily a bad idea.

At the same time, the agricultural system was changing and the Station was not always quick to grasp the importance of its alteration. Some of the changes resulted from the actions of government. Federal crop loan programs initiated during the depression imparted an unprecedented degree of price stability to wheat and other basic commodities. These programs were designed to be temporary, but by the late forties they were assuming most of the aspects of permanence. Federal crop insurance, another product of the New Deal, mitigated some of the natural threats producers faced. Such programs went a long way toward addressing the problem of income stabilization, diminishing the necessity of diversification in the process. Indeed, diversification could even become a problem when it drained away the farmer's time, energy, and capital. Federal lending programs and farm cooperatives further diminished the farmer's risk. Then, too, the farmers were different. They were better educated, more sophisticated, and more flexible than ever before. Because they were better farmers and businessmen they narrowed the risks their fathers had faced. The tools they had also minimized their risk. Chemicals, machinery, and more and better adapted varieties of more crops left them freer from natural calamities than their counterparts in an earlier generation.

Modernity in agriculture was paralleled by modernity in rural living. North Dakota farmers were increasingly inclined to send their children to high school or even college. Improved roads and automobiles made towns and cities more accessible. More than ever before rural North Dakotans looked, dressed, talked, and acted like their brothers and sisters in town. Electricity had a tremendous impact on rural life. In 1945 only one in ten North Dakota farms was connected to a central power station, despite the fact that the Rural Electrification Administration had existed for a decade. By 1951 three out of every four of the state's farms were hooked up. The result was a dramatic narrowing of the gap between rural and urban life as "most families acquired radios, electrically driven washing machines, . . . irons, . . . refrigerators, . . . toasters and clocks . . . [and] vacuum cleaners."[28]

The structure of agriculture was also changing. As early as 1941 Walster could see a challenging future: "The State must choose which highway she is to follow. Is this State to be one of many relatively small farms with many farmers on the land, or is it to be one of a few large farms with few farmers on the land? The issue cannot be dodged, it must be faced." The state seemed firmly in favor of the small farm. It had outlawed corporate ownership of land in 1932 and seriously debated instituting a graduated land tax in the forties and early fifties. The Station favored the small farm as well. Rainer Schickele was an especially eloquent exponent of the small, or "family," farm. He saw the family farm as the superior of the corporate farm not only in its productive efficiency but also in the moral values it represented and inculcated. "The Jeffersonian ideals of a democracy based upon an agrarian society of independent owner-operated family farms has become outdated," Schickele admitted in 1948, "but his [Jefferson's] vision of the cultivators of the earth as the most valuable, vigorous, independent and virtuous citizens was as valid then as it is now . . . ." He concluded that "there is no ground more fertile for the growth of democratic attitudes and institutions than a community of reasonably well-to-do and secure farm families."[29]

Developments during the forties and fifties dramatically changed the definition of a family farm in North Dakota. Modern machinery made it possible for a family to farm more than a quarter section or two, and the desire to enjoy a modern standard of living made it necessary. Station economists noted in 1952 that "larger grain farms are more efficient than smaller farms" and that "economies of scale are to be had at least up to the 2-section size of farm." They admitted that "since 1930, North Dakota has lost one-third of its farm population" but argued that as a result there existed "a better balance between people and land" and improved "living levels and economic opportunities over the state." Still, there were other effects of the changing structure of agriculture that went beyond the fencerows. Smaller rural population made the maintenance of schools, churches, and public services more difficult. Fewer farmers meant fewer customers for town merchants. As the capital requirements of agriculture increased, entry into farming became more difficult.[30]

How big was too big? How few were too few? What should be done and how? These were questions asked in state and Station after World War II. Agricultural economist Baldur Kristjanson explored the use of farm credit and property tax assessment as means of keeping farms small. He was also an advocate of the graduated land tax proposal pressed by the North Dakota Farmers Union.[31]

Station investigation of the social implications of structural changes in agriculture ended rather abruptly in 1954 and 1955. Rainer Schickele resigned to take another position, and Baldur Kristjanson and Cecil Haver, another agricultural economist, were fired after an acrimonious struggle over faculty governance with College president Fred Hultz. Their successors in Agricultural Economics directed their research to issues such as marketing and adoption of innovations. While they chronicled the continuing structural alterations in North Dakota agriculture, they tended to ignore the normative issues Schickele and his colleagues had explored.[32]

The Station observed the changes in agriculture after World War II, but it also participated in them. The chemical revolution initiated during the war was joined eagerly by the NDAES. Chemical fertilization—especially the use of nitrogen—promised to help North Dakota farmers narrow the gap in yield that separated them from producers in more favored climates. Soil scientist Joseph Zubriski, who joined the staff in 1950, believes the "100-fold increase in the use of [chemical] fertilizer since World War II" to be the major reason for the dramatic yield increases of the past few decades. New chemicals for the control of such chronic problem weeds as wild oats, pigeon grass, and wild mustard promised to increase yields further.[33]

Chemical pest control was equally promising. Entomologists who had struggled for years to control such destructive insects as grasshoppers, sawflies, and potato bugs with such primitive tools as tar pans, hopper dozers, and arsenic and kerosene compounds eagerly grasped new chemical weapons. The most important of these was DDT. As station veteran John Callenbach remembers, "DDT . . . revolutionized everything . . . entomology really took off." DDT was the magic bullet that promised even to eradicate whole species of insect pests. Certainly, no more effective means of control had ever been devised. The chemical companies pushed it and the farmers welcomed it. "Growers were very supportive," entomologist Richard Post remembers , and "the chemical companies just gave us this stuff, they didn't worry about it."[34]

As is the case with most apparently progressive steps, the chemical revolution was not without its costs. Many farmers wanted to increase their yields, but greater production exacerbated the chronic oversupply problem in American agriculture. Moreover, chemicals drove farm costs higher and intensified the cost-price squeeze. Overused nitrogen ran into surface waters and seeped into ground water supplies, threatening the health of people, livestock, and wild animals. Herbicides carried health risks about which few, if any, researchers or users were aware. Such insecticides as DDT and chlorodane promised unprecedented control, but at a price. These compounds wiped out benign and malign insects indiscriminately and carried health risks to humans and animals as well. None of this was readily apparent. "Nobody could foresee the environmental problems which have come from chemicals," John Callenbach points out. "We did not consciously harm the environment for commercial purposes."[35]

One aspect of chemical insecticide usage that troubled producers and researchers early on was the development of immunity by targeted pests. Insects are marvelous creatures. They preceded humans on the planet and most entomologists believe they will be here long after we are gone. Rather than eradicating them, DDT and the other wonder compounds merely presented them with a special challenge, and they were not long in overcoming it, much to the consternation of farmers, researchers, and chemical companies. Imagine Dick Post's alarm when he discovered in 1952 that some Colorado potato beetles had developed a resistance to DDT.[36]

Having placed all of their eggs in the alluring chemical basket, entomologists were ill-prepared to confront such problems as environmental degradation and insect resistance. Fortunately, one NDAES researcher, veteran entomologist J. A. Munro, was still looking backward. He had worked for years on biological and cultural control of insects, and he did not rush to jump abroad the chemical bandwagon. While conceding in 1951 that "the most common method of controlling insects is by chemical means," he contended that "this method . . . must be supplemented wherever practicable by . . . cultural and biological control." Munro's work on biological control of the corn borer and the sweet clover weevil provided a bridge between the old, pre-chemical entomology and the new integrated pest management that de-emphasizes chemical use. While many other stations were, in effect, reinventing the wheel, the NDAES had a tradition of biological control work on which to draw.[37]

IV

In such traditional areas of research concentration as agronomy and entomology the relatively flush times of the forties and fifties allowed the Station to upgrade staff and facilities. Farm groups were generally supportive of Station programs and the Board of Higher Education and the legislature were relatively liberal with funds. The problem was that most new resources had to be devoted to repairing the damage of a quarter century of low appropriations.

Physical facilities called for immediate attention. Dick Post remembers that Entomology had "a pitiful greenhouse—kind of a lean-to" when he came in 1946. Animal scientist Bill Dinusson recalls that "facilities were damned few. We had one building for sheep—nothing for hogs. There were no research facilities for beef cattle." Even Agronomy, the relatively favored department, had "facilities [that were] not so good" to the thinking of a young Jack Carter. President Hultz hoped to leave a legacy of buildings. In order to achieve his goal he insistently pressed the legislature for funds. Hultz was not always successful, but substantial renovation and upgrading of existing facilities was accomplished, and some new structures were built. Van Es, where Veterinary Science is located, was substantially remodeled, Shepperd Arena was completed in 1952, and plans were made for a new home for Agronomy and related plant sciences. In 1959 these were realized with the completion of Walster Hall. Adding to the comfort of all on campus was the addition of concrete sidewalks and the paving of streets.[38]

The postwar years also witnessed substantial hiring. The Station had been understaffed for years. "These oldtimers . . . were marvelous people, but they couldn't do everything," Glenn Smith notes. "We didn't have enough to fill in everything." After World War II the Station began to fill in, adding researchers and administrators who would lead the NDAES for a generation. Animal scientists Buck Buchanan, Bill Dinusson, and Merle Light, agronomist Jack Carter, entomologists John Callenbach and Dick Post, soil scientists Enoch Norum and Joe Zabriski, bacteriologist John Doubly, chemist Harold Klosterman, horticulturist Robert Johansen, and administrators Arlon Hazen and Peder Nystuen were some of the many new people who came in the immediate postwar years.[39]

The new personnel did more than simply fill gaps. They also helped upgrade the Station as a scientific institution. The NDAES was fortunate in being able to hire at a time when there was an explosion of Ph.D.'s in the sciences. Young men and women, inspired by the wonders of science and benefiting, in many cases, from the G.I. Bill of Rights, flooded in and out of graduate schools. The resulting buyers' market was a bonanza for employers such as the NDAES, allowing them to attract quality researchers who would not have given them a second look just a few years before. The influx of young, well-trained scientists was a mixed blessing, however. There were inevitably some frictions with station veterans. The new people were confident—sometimes even brash—and they knew little about the Station or the state. Most came with the intention of moving quickly to something better rather than devoting their careers to serving North Dakota agriculture. The "good young people . . . saw North Dakota as a stepping-stone," Glenn Smith recalls, "but some of them stayed." Those who stayed enhanced the Station as a scientific institution and made enduring contributions to the agriculture of the state.[40]

The improved position of the Station was reflected in the comments of USDA inspectors. In 1951 an investigator from the Office of Experiment Stations wrote Walster that "the various Departments are . . . well staffed . . . [and] the various research programs appeared to me to be well-suited to the agricultural needs of the State . . . and I was pleased to note a good balance of practical and fundamental research in many instances." The next year, another inspector praised the Station for high staff morale and "very fine results" in the research program.[41]

The NDAES upgraded after World War II, but it did not expand very much into new areas. Relatively prosperous economic times still did not result in enough resources to allow the Station to do everything it wished to. "We have certain lines of activity not represented at all in this Experiment Station which certainly need building into our scheme of things," Walster admitted to Director Trullinger of the OES in 1952, "notably home economics investigations . . . forestry investigations . . . and utilization investigations, from the point of view of non-agricultural uses." To a Farmers Union official interested in home economics investigations Walster wrote that the Station could not undertake such research without "*added personnel* . . . and a place to do the work." The resources that came during the postwar years were sufficient to allow the Station to do traditional things better. They were not sufficient to allow expansion into neglected areas.[42]

The years after World War II witnessed a changing of the guard at the Station. New researchers came in while the veterans of an earlier era retired or died. Agricultural economist Cap Miller, agronomists O. O. Churchill, Olaf Heggeness, and L. R. Waldron, botanist H. L. Bolley, dairy specialist J. R. Dice, poultryman O. A. Barton, and bacteriologist Casper Nelson all passed from the scene after the war. Bolley had been with the Station since its founding, and some of the others had worked there since early in the century. They were men who had shared the struggle to maintain the institution and serve North Dakota agriculture in the face of tremendous adversity. They were shaped by the past and they looked to the past, but because of their efforts a promising future beckoned. They were men of faith and loyalty and the state they served was richer for their presence.

What was true of the staff was even truer of the director. H. L. Walster retired in 1953 after thirty-four years of service. The leave taking of the Wisconsin-born scientist who became "almost as much a part of North Dakota as the Missouri River" was a gracious one. From throughout the state came a genuine outpouring of affection for this devoted servant of North Dakota. Walster had always attempted to maintain a scientific institution of the highest possible quality in the face of great obstacles. He had constantly stressed the importance of scientific rigor and teamwork. He had helped repair the Station politically after the disastrous thirties and prepare it scientifically for the second half of the century. He had been a visible and active director, carrying the Station's message into every corner of the state he loved. He always told North Dakotans the truth, whether they wanted to hear it or not; it was not in his character to do otherwise. Walster saw himself as the lucky partner when he reflected on his long relationship with the Station and the state, noting that "I have the most interesting job in North Dakota." It is pleasant to think that Walster benefited from his years at the Station, but it is hard to believe that his rewards could have even approached his contributions.[43]

8

THE GOLDEN AGE

Within the past 15 years the Agricultural Experiment Station has about doubled in size in terms of budget,personnel and activities. Even with an inflationary factor applied to the budgets, this growth has been noteworthy. We expect to continue to grow at the rate of 10 to 20 percent into the immediate future to take care of the rapidly increasing needs of North Dakota agricultural industry for basic knowledge and interpretation.

Arlon G. Hazen, 1968

An experiment station belongs to the people, the more they use it, the better it serves them.

Author Unknown

The quarter century following the retirement of Director H. L. Walster in 1954 was a golden age for the North Dakota Agricultural Experiment Station. It was a golden age in terms of dramatically increased financial support, as Director Arlon G. Hazen noted in 1968. But it was also a golden age in the sense that the Station was able to approach its potential in serving North Dakota agriculture. It was able to do more than ever before, more farmers used it than ever before, and it was more genuinely appreciated than ever before. As an unidentified author sagely noted in *North Dakota Farm Research* in 1963, "an experiment station belongs to the people, the more they use it, the better it serves them."[1]

The fortunes of the NDAES reflected the improved position of North Dakota agriculture during the fifties, sixties, and seventies. Agriculture continued to be the main generator of new wealth in the state, and it enjoyed modest prosperity through most of this period. Generally favorable growing conditions, technical developments in agriculture, and government price support programs imparted an unprecedented degree of stability to farming and farm incomes.

North Dakota was becoming a more important contributor to the nation's agriculture. In 1954 the state marketed $376 million in agricultural products, ranking it twenty-fifth in a forth-eight-state nation. Twenty years later it stood seventeenth in a fifty-state nation, marketing products valued at over $1.8 billion. Part of this change resulted from farm abandonment, suburbanization, and reforestation in other states. But part of it also resulted from the positive actions of increasingly innovative and sophisticated North Dakota farmers.[2]

Dramatic changes in the structure of agriculture and in the farming business occurred during these years. Government programs rewarded farmers who brought in heavy yields, and modern equipment and chemicals helped make huge crops commonplace. At the same time, rising input prices squeezed farmers and forced them to strive for greater economies. The growing capital requirements of agriculture and the technological multiplication of individual productivity seemed to dictate that farmers achieve economies of scale through expansion. The result was fewer, larger, and more heavily capitalized farms. In 1954 there were still 62,500 farms in North Dakota, averaging just over 680 acres each. Twenty years later there were only 43,000, and by 1979 barely 40,000. In that year the average farm had over 1,000 acres. Farming in North Dakota was becoming a bigger business practiced by increasingly sophisticated people.[3]

Business skill and technical competence were not the only areas where the greater sophistication of farmers could be seen. The North Dakota farmer of the golden age was unwilling to live as his or her parents had lived. Farm families demanded modern homes with water and electricity and associated conveniences. They wanted to travel on improved roads to distant centers to shop, they wanted winter vacations, and they wanted their children to receive college educations. They were less likely than their parents to look backward at Old World deprivations and New World dust and depression, and more likely to look to brighter tomorrows.

The NDAES was ready and eager to help North Dakota farm people realize their bright tomorrows. It could offer advice on new crops and better varieties of old ones. It could suggest optimum fertilizer and pesticide usage. It could help farmers better feed livestock and care for the soil. And it could provide up-to-date management tools for the farm business. In short, the Station could, and did, offer an expanding array of services related to production agriculture, and eager North Dakota farmers took much of what was offered.

The Station prospered because it could serve farmers' needs, but that was not the only reason. The NDAES in the golden age benefited from what it had done before. Walster and his staff had planted the harvest that the Station reaped in the golden age. They had proved that dedication and a commitment to service were best for the state, even if those qualities were not always in season.

The Station also benefited in the golden age from its strong political position. No major organization in the state opposed the NDAES, and most worked for it. The increasing importance of commodity groups in North Dakota agriculture resulted in new and effective support for the NDAES in the legislature.

And the Station was careful to cultivate key legislative leaders, elected and appointed members of the executive branch, and the members of the state's congressional delegation. North Dakota politics had been a school staffed by hard teachers, and the Station had learned its lessons well.

The changing environment in which North Dakota agriculture operated, the previous services of the NDAES to the state, and the political position of the Station are all necessary components in explaining why the fifties, sixties, and seventies were a golden age, but they are not sufficient. People were also instrumental in helping the Station reach its potential, and none was more important than Arlon Hazen, who directed the NDAES from 1956 until 1978. "North Dakota was ready for and needed Arlon Hazen," agronomist Jack Carter contended, because Hazen helped shake North Dakota agriculture out of its depression mentality and point it toward the future. Perhaps others could have done the same thing, but because Hazen was the key figure in this endeavor he dominates the story of the golden age.[4]

The characterization of the fifties, sixties, and seventies as a golden age should not lead readers to conclude the Station had no difficulties. There were occasional budget cuts, inflation was an increasingly frustrating problem, and there was a fundamental challenge to the Station's commitment to quality wheat. By the seventies the agricultural research enterprise was once again under assault in the United States, and there were echoes of that in North Dakota. But in light of the Station's difficult past and uncertain future, the quarter century after 1954 can fairly be called a golden age.

I

When H. L. Walster retired as director of the NDAES in 1954 he was replaced by veterinarian Glenn Holm. Many within the Station favored Associate Director Glenn Smith for the post, but, perhaps because Smith and President Fred Hultz "didn't see eye-to-eye," Holm was made Walster's successor. Holm was well regarded and highly respected while he was director, but he left after only two years to head a new college of veterinary medicine at Oklahoma A and M.[5]

Holm was succeeded by Associate Director Arlon G. Hazen, an Oklahoma native who had a master's degree in agricultural engineering from Iowa State. After service with the Army Corps of Engineers during World War II, in which he attained the rank of lieutenant colonel, Hazen came to work for the NDAES. He served as superintendent of the branch station at Williston before coming to the main station to assume the post of associate director.

Hazen brought significant administrative talents to the Station. He was extremely efficient—Leroy Schaffner contends "he was the most efficient administrator I've seen." Known for his care and deliberation, he never made decisions in haste or without weighing all relevant points of view. "He was meticulous,"

David Ebeltoft remembers, "he never shot from the hip." Once he made a decision he could be depended on to stick to it.[6]

Two of Hazen's most respected traits were his honesty and his loyalty to his staff. Harold Klosterman is but one of the staffers who remembers him as a "square shooter." Dickinson superintendent Tom Conlon remembers that he "stood behind his people 100 percent. He might privately give you hell, but publicly he was 100 percent behind you."[7]

Hazen's major handicap was his lack of a Ph.D. degree. When he went to school the master's was the terminal degree in agricultural engineering. When universities began offering Ph.D.'s in agricultural engineering after World War II Hazen and others in his situation appeared undereducated. He compensated for his dearth of formal training by careful and determined study. In essence, he made himself an expert in all of the disciplines represented at the NDAES. "He was a quick learner," Glenn Smith remembers. John Callenbach, one of Hazen's associate directors, adds "you didn't have to draw him any pictures."[8]

Some of the traits that served Hazen well at the Station also made him a popular and effective public figure in the state. He lacked Walster's impressive erudition, but he related well to farmers because he understood them and their problems so well. He dealt very effectively with legislators and other politicians on an individual level. Stanley Moore of the Farmers Union remembers him as "an excellent politician—very adroit." Although he was sometimes seen as a wizard in dealing with the legislature, Hazen's success was not based on magic. As Art Link, who related to Hazen as a legislator, congressman, and governor, wisely notes, Hazen's political successes came because "he knew his topic, did his homework, and was honest."[9]

Hazen's political abilities were a significant asset to the NDAES. Not only did the Station continue to depend on legislative appropriations, it also had to adjust to fundamental shifts in the political environment of North Dakota agriculture during the fifties and sixties. During these years commodity groups rose in importance and power while general farm organizations declined. This development had important implications for the NDAES.

Commodity groups—organizations of producers of a single agricultural commodity—had existed in American agriculture since the late nineteenth century and had grown dramatically in the 1930s, partially stimulated by commodity-oriented New Deal farm programs. Developments that occur elsewhere often are delayed in coming to North Dakota, and such was the case with commodity groups. With the exception of a few types of farmers, such as dairymen and potato growers, most North Dakota producers of particular commodities did not create durable organizations until the mid-fifties. Between then and the mid-sixties all major and most minor commodities came to be represented by organized groups.

Commodity groups attracted farmers for a number of reasons. Their relatively small size and narrow focus made it easy for them to maintain unity and helped make them effective lobbying organizations. Because they were concerned mainly with production and marketing problems pertaining to particular commodities they were able to bridge the philosophical gaps that separated farmers in North Dakota and elsewhere. Finally, unlike such general farm organizations as the Farmers Union and the Farm Bureau, commodity groups were not compelled to engage in the often-divisive struggles that accompanied the shaping of national farm programs. As Leeds farmer Al Kenner, who was instrumental in organizing both the Farm Bureau and the durum growers in North Dakota, notes, "I could make better use of my time with the commodity groups than with the Farm Bureau. Commodity groups are interested in promotion and marketing—we don't want to get involved in farm programs."[10]

In its early years the NDAES had been active in organizing dairymen and grain growers—the former successfully and the latter unsuccessfully. While much had changed since the days of Ladd and Worst, Station support for commodity based organization in agriculture never wavered. The Station favored commodity groups because they included active and interested growers who could provide scientists with information about problems and through whom researchers could work. The Station also perceived, correctly, that commodity groups would be more devoted supporters of research than were general farm organizations and that they would lobby the legislature effectively for funds. Finally, the Station saw these organizations as private sources of funds for projects on particular commodity problems. In order to take better advantage of what they had to offer and to better serve them, the Station drew on commodity groups to form "advisory committees" on a variety of crops and livestock products. The stated purpose of the advisory committees was to "assist in defining the goals and objectives for agricultural research and services provided by the University and, in general, to improve and enhance the effectiveness and image of the University and its agricultural teaching, research and extension programs."[11]

Commodity groups have generally met the positive expectations of the Station. They provide an interested and sophisticated clientele with which researchers can work, and they are strong supporters of legislative appropriations. "If we hadn't had commodity groups we'd be slipping around with about seven faculty members," plant pathologist Richard Kiesling notes. "Commodity groups get the legislature's attention because they speak the language of votes." Direct financial support from commodity groups has also been substantial. Of the Red River Valley Potato Growers, for example, Robert Johansen says, "if I didn't get support from them, I'd have to close the doors."[12]

The growing dependency of the NDAES on commodity group support has had its unfortunate aspects. Station scientists must do a balancing act that is sometimes difficult, fulfilling the expectations of commodity groups while maintaining control over their research agenda. Commodity groups' interest in solutions to immediate problems strengthens the traditional Station concentration on applied research and makes it difficult to shift resources to more basic work. Finally, the relationships commodity groups develop with particular departments contribute to a sort of Balkanization in the Station and result in more rewards going to some segments of what is supposed to be an interdependent entity than to others. Speaking of 1987 budget cuts, biochemist Harold Klosterman notes, that "we got clipped . . . because we don't have the support of any commodity groups . . . . Botany, biochemistry, and microbiology have all fared poorly in this environment." These disciplines tend to be more oriented than others to basic research; their fate underscores the heavy stress on practical work the rise of commodity groups has brought.[13]

On balance, though, the Station has benefited from the rise of commodity groups and its close relationship with them. And that relationship will probably become closer. No less an observer of North Dakota agriculture and public affairs than Governor George Sinner believes that, in the face of tight government resources, the Station must become more dependent on commodity group support. "Private sector investment has grown, is a good direction, and is the wave of the future," Sinner concludes.[14]

The rise of commodity groups added another set of relationships to what was already a complicated political matrix. Hazen proved well qualified to handle all of the Station's relationships deftly. He continued to maintain close relations with such agribusiness groups as the American Malting Barley Association and the Crop Quality Council, an organization formed in 1960 by the merger of the Northwest Crop Improvement Association and the Rust Prevention Association. The Garrison Diversion Conservancy District, the Farmers Union, and the Greater North Dakota Association were all firm supporters of the Station. Hazen was especially sedulous in fostering close ties with the GNDA, and he succeeded in that effort. Writing GNDA president R. M. Stout, Jr., in 1975, Hazen thanked the group for being "an ardent supporter of our pleas for additional resources to more nearly meet the almost unlimited research . . . potentials of North Dakota. These efforts and interest have undoubtedly been instrumental in our financial considerations by the legislative sessions during these years."[15]

Hazen also maintained good relations with elected officials in the state and national legislative and executive branches, concentrating always on those in key positions but never ignoring anyone. He worked particularly closely with Milton Young, "Mr. Wheat," whose seat on the Senate Appropriations Committee potentially conveyed great benefits to the Station. Hazen and Young, who were on an "Arlon" and "Milt" basis, obviously had great respect for one another. Hazen wrote the senior senator in 1966, "I doubt there are many, if any, other members of Congress who have as fully understood and consistently supported this very vital program as you have so effectively done." In 1971 Young wrote, "Arlon, it has been a pleasure for me to work for the badly needed funds for this program over the years. I do feel very strongly that the investment we make in agricultural research through our State Experiment Stations is a good one and one that must be continued."[16]

The political strength of the Station and Hazen's adroitness in using it were frequently demonstrated during the sixties and seventies. One goal of the NDAES and its friends was realized in 1964, when the voters approved a constitutional measure separating the legislative budgets of the Station and the Extension Service from those of the colleges and universities. The Station had argued, convincingly, that this change was necessary because agricultural research was a unique and critical public function that could not be effectively funded using what was essentially an enrollment-driven formula.

Another important change came in 1972, when the State Board of Higher Education agreed to create a special board to consult it on matters related to the Station and Extension. The SBHE represented a vast improvement over the highly politicized Board of Administration that preceded it. It was not, however, a perfect instrument for setting agricultural research policies and priorities. Its members were—and are—drawn from many parts of the state and walks of life. Sometimes some of them know something about agricultural research and sometimes not. This situation, which was a constant irritant to the Station and its champions, came to a head in 1972 when Casselton farmer George Sinner joined the Board. Sinner remembers that "as a farmer, I was expected to know everything about extension and research." Sinner refused to bear this burden, suggesting instead that a consulting board be created. He was strongly supported by Hazen and NDSU president L. D. Loftsgard as well as by a number of farm legislators. The SBHE went along with Sinner's suggestion.[17]

The new body was left largely to define its own mandate, a process that resulted in some creativity and also some strife. In practice, the Consulting Board tried to be "a catalyst between the Station, Extension, . . . the commodity groups," and the SBHE, as early member Lois Jones remembers. It has continued to function in that same general way. The board meets representatives of commodity groups for their suggestions and receives presentations from and makes investigations of the Station—at times urging "changes in direction and an end to some research," as former member Keith Bjerke notes. It then makes suggestions regarding funding and policy to the SBHE.[18]

The process has been marred by some frictions and frustrations, especially in the early years. Former governor William Guy, who belonged to the first Consulting Board, believes the members "always had a problem with realizing where their authority began and ended" and thus behaved presumptuously at times. For its part, the State Board of Higher Education was not always receptive to its advisory body and sometimes even seemed resentful. Perhaps SBHE members saw the consultants as raiders on the treasury or feared they would encroach on board authority. Then there was—and is—the problem inherent in any consulting body that it can advise but cannot execute. As onetime Consulting Board member Kelly Miller puts it, "there's a woman here in Towner who has an old tom cat that used to go out at night and came back all beat up, so she had him fixed. He still goes out at night, but now he just consults."[19]

The Consulting Board has not been perfect from anybody's standpoint, and there are those—former GNDA official and legislator Clark Jenkins and leading cattleman Jack Dahl, for instance—who do not believe the State Board of Higher Education should have any control over research and extension. Still, it provides a means that did not exist previously for commodity groups and researchers to have some say in the setting of budgets and policies and for SBHE members to learn something about the research system and its needs.[20]

II

Recognition of the political environment in which the Station operated during the fifties, sixties, and seventies is essential, but it should not cause us to ignore a key fact: The NDAES enjoyed public support not because of certain favorable political developments, but because it served North Dakota agriculture through its research programs.

Those who knew him agree that Arlon Hazen was a very conservative man in every way. That conservatism was reflected in the research program over which he presided. Research was heavily oriented toward applied, production-oriented work in crops. In 1957 incoming director Hazen explained to county agents what he conceived the mission of the Station to be: "Our program is largely one of trying to keep up with current agricultural problems such as animal and crop diseases, insect pests and a maintenance of our soil at a profitable level of production. In addition, effort is directed toward lowering of unit costs of production for the farm family." Hazen conceived of basic research as a luxury the NDAES could ill afford, telling the agents that "the reason you do not find more basic or fundamental research projects is simply because we have neither the funds nor the personnel to assign to this kind of work in face of current pressing problems which are largely in the area of applied research." Hazen was so oriented to applied research that he contrived to define all research in those terms. For instance, he complained of the USDA's definition of marketing research, arguing that "the marketing definition should apply to any research which has as an objective an improvement in the marketing or marketability of any product . . . . For example, if a potato is known or thought to sell better on the market due to a certain skin color or a certain shape, then it is marketing to develop and make available such a potato."[21]

In Hazen's lifetime the emphasis on applied, production-oriented research came to be questioned nationally. Nevertheless, this emphasis was supported by two compelling realities. First, it was firmly within the tradition of the NDAES. The Station had always emphasized applied work on standard North Dakota crops. Given the expense of basic research and the Station's usual spare budgets, that emphasis had always seemed best designed to serve farmers' needs. Second, most farmers were most concerned most of the time with production problems, for which they wanted quick solutions. The state and national agricultural environment during Hazen's directorship, which emphasized high yields, producer efficiency, and economies of scale, strengthened rather than weakened North Dakota farmers' orientation toward immediately practical applied work.

Within the research program the Station during the Hazen years continued to be devoted largely to crops, especially grains. Answering an inquiry from a farm paper in 1977, Hazen noted that within the Station budget "we devote approximately 31% to cash grains; 26% to range and livestock; 20% to natural resources [soil and water, especially]; 13% to weeds, seeds, disease, insects; 6% to horticulture; and 4% to farm and community development."[22]

Small grains research, always important at the Station, thrived and expanded in the favorable environment of the Hazen years. The breeding of new varieties appropriate to the area remained the central concern in Agronomy. The fifties, sixties, and seventies saw the introduction of several new lines of barley, durum, hard red spring wheat, and other grains as well. The main hard red spring introduction was Waldron, released in 1969. Waldron was a high-quality variety and a dependable yielder, but it tended to be susceptible to ergot, a fungal disease. Leeds and Wells were the major durum introductions during the Hazen years. Farmers' faith in the NDAES was reflected in the fact that "Waldron constituted 62% of the hard red spring wheat acreage in the 1971 crop year," while Leeds and Wells made up nearly 98 percent of the durum acreage.[23]

The breeding program was what the Station was best known for, and it was the source of much public support as well as needed funds from commodity and industry groups. It was important for the NDAES to remain on the cutting edge in agronomy, which necessarily meant carrying on some basic research. The main thrust in fundamental research in agronomy initiated when Hazen was director was an ambitious effort to develop a hybrid wheat that would allow yields to be boosted dramatically. Though that long project has not succeeded in creating a commercially feasible hybrid, it "has helped make NDSU one of the two or three outstanding wheat research centers in the world."[24]

Breeding new varieties was one matter, but getting them to North Dakota farmers quickly enough and in large enough quantities to provide an economic advantage was quite another. The North Dakota Seedstocks Project, which was created as a formal entity in 1961, represented an attempt to attack the twin problems of increase and distribution.

The Seedstocks Project involved the systemization and coordination of individual practices that had been followed for many years. It provided for the rapid, controlled increase of promising varieties through the coordinated use of greenhouses, winter acreage in Arizona, California, and Mexico, Station acreage, especially at the Agronomy Seed Farm, and the acreage of cooperating seed growers. When each segment of the Seedstocks Project worked at optimum efficiency the results could be stunning. For example, in 1959 ten pounds of Trophy barley were sent to Mexico for winter increase. By 1963, enough Trophy was available to seed 42 percent of the state's barley acres and produce thirty-nine million bushels.[25]

The Seedstocks Project was not free of controversy. Because the Station profited from the sale of substantial quantities of seed for new and desirable varieties, representatives of the seed trade and even some certified seed growers complained of unfair competition from a publicly supported entity. For example, Richard Crockett of the GNDA complained that the NDAES was "trying to make commercial farms out of experiment stations" and noted that Agronomy chair Jack "Carter became the biggest seeds man in North Dakota." But the main purpose of the project was to serve the farmers of North Dakota, and its champions argued that it fulfilled that function in a splendid manner. In 1971, for example, the early introduction of Waldron put an extra fourteen million dollars in the pockets of North Dakota farmers. Because it is profitable to both Station and clients, it is no wonder that the Seedstocks Project "is the envy of a great many states," in David Ebeltoft's words.[26]

Among the agronomists and cereal technologists at the NDAES the emphasis remained on high-quality wheat as opposed to high-yielding wheat. In a dream world—or in a magical hybrid variety—quality and quantity might go together, but in the real world they do not. Quality wheats—those with high protein, high gluten content, high ability to absorb water, and other favorable baking characteristics—do not yield heavily per acre. Wheats that produce heavy yields, on the other hand, are relatively deficient in quality.

The Station had sound reasons for stressing quality over quantity. With its short growing seasons, cool nights, and relative aridity, North Dakota provided an ideal environment for high quality wheats. These same factors were among those that made it impossible for North Dakota to think seriously about yield competition with the more favorably situated winter wheat states of the Central Plains. As Hazen explained in a letter to Kent Jones, then a North Dakota legislator and later state commissioner of agriculture, "there is little . . . real competition for the *quality* which has been maintained in our hard red spring and durum wheats....There is an endless competition for *quantity*, and we are on the losing end of this competition for two major reasons—distance to market and cost of production (a reflection of our lower yield per acre due primarily to shorter season)."[27]

Several factors helped producers of quality wheats compensate for their lower yields. Because quality wheats were in demand for blending in bread making, they moved easily into national and international channels of trade and they usually commanded a price premium. While that premium was never as high as producers thought they deserved, agricultural economist Don Anderson concluded North Dakota wheat sold for an average of twenty-two cents more per bushel between 1962 and 1967 than did Kansas wheat, and thirty cents more than Illinois wheat. In order to encourage the maintenance of wheat quality the Crop Quality Council conducted extensive tests every year on the basis of which it established a list of "approved" varieties to guide farmers and buyers. The United States Department of Agriculture further encouraged quality wheats by discounting what it considered to be low quality varieties when making commodity loans.[28]

The elaborate edifice supporting wheat quality began to crumble in the late sixties. Private seed companies introduced and aggressively marketed semidwarf varieties appropriate to the Northern Plains. Semidwarf wheats, developed in Mexico by Nobel Prize winner Norman Borlaug and his associates, matured rapidly and returned high yields in dry climates. To farmers squeezed between costs and prices these high-yielding varieties were understandably attractive. To the NDAES and the Crop Quality Council the semidwarfs were threatening because they lacked the quality of such varieties as Waldron. Hazen considered the specter of the semidwarfs to be an example of creating a long-term problem by trying to solve a short-term one. He noted that a shift to semidwarfs might provide "greater income at the moment. Yet, we may well be setting the stage for an even greater future problem of having disrupted or destroyed a market which depends heavily upon consistent maintenance of quality."[29]

The grain trade provided little support for the Station in its struggle against the semidwarfs. While the Crop Quality Council found the semidwarfs generally inferior in quality, it decided in December of 1969 not to label them unacceptable when it was threatened with an antitrust suit by World Seeds, one of the companies marketing semidwarf varieties. World's attorneys argued that the approved list implicitly restrained trade. Perhaps fearing similar legal action, the Closing Committee of the Minneapolis Grain Exchange removed premium payments for high-quality wheats less than a month later.[30]

The USDA also backed away from supporting quality. Eager to remove government controls from agriculture and encourage private sector initiative, the Nixon administration began talking about dropping the "undesirable wheat variety list" and ending the system of discounting loans on such varieties. Despite protests by millers, bakers, some stations, and some producers, the USDA dropped the list effective with the 1973 crop year. "In line with USDA policy of encouraging producers to place more reliance on the marketplace and because of numerous varieties being released by private breeders," the USDA concluded, "a more effective way of discouraging low quality wheat is for buyers to continue to apply discounts in the marketplace and for crop improvement organizations to continue promotion of quality wheat production." Whether that was feasible in light of the antitrust threat remained to be seen.[31]

The NDAES continued to battle for quality wheats by urging growers not to raise semidwarfs, stimulating the opposition of seed companies and some producers in the process. Ken Gilles, vice-president for agriculture at NDSU and a trained cereal technologist, complained to an executive of the Minneapolis Grain Exchange that "one of the private seed companies is using the theme that the farmer is a forgotten man and the inference is that he has been forgotten by the

State Agricultural Experiment Station." Farmers accused the Station of jealously opposing new varieties and of favoring Station releases because of pride rather than concern for the wheat growers. One state legislator and farmer from McCanna, for example, wrote that "many of my neighbors and I are not so convinced of your unbiased opinion that we are willing to take it all for fact just because you say so," and threatened that "the private seed companies may take over and your secure jobs may be lost." Arden Burbidge, a prominent Park River farmer, complained of "a general philosophy of paternalism on the part of experiment stations . . . which . . . defeats the objective of increasing farm income from wheat in North Dakota."[32]

Such accusations were painful to those who saw themselves as servants of North Dakota agriculture. Hazen assured the readers of *North Dakota Farm Research* that "we expect and encourage our producers to make their own individual variety selections." He added that Station recommendations were "human judgments based upon experience, available information and a dedicated desire to help maintain and improve the economic and social well-being of North Dakota people." Stung by the charge of paternalism, he assured President Loftsgard that "if there is a paternalistic philosophy and attitude within the agricultural experiment station, it is in the direction of being paternal for our producers." That did not really respond to the point Burbidge made, but it did indicate Hazen's chagrin that the NDAES might be seen as self-serving.[33]

During the controversy over the semidwarfs the Station stood accused of unworthy motives, but it fought the new varieties because it recognized—as did researchers in the federal government and the milling and baking industries—that they were deficient in quality. Tests in 1971, for example, showed that "none of the semidwarfs tested . . . met the acceptability standard . . . . All appeared deficient at least in the following three major quality categories: low protein, low absorption, [and] excessive mixing requirements." Nor was low overall quality the only problem. The semidwarfs also exhibited a disturbing unpredictability in quality, varying greatly from place to place and even from field to field. Writing to a seed company official, Hazen expressed his concern that "the semidwarf wheats which we have seen grown in North Dakota so far exhibit a much greater *range of variability* (unpredictability) than do the so-called 'standard varieties.'" The threat semidwarfs presented was that the state would lose its reputation for quality by engaging in a yield battle it could not win. As Jack Carter put it, "an increase of HRS [hard red spring] of unacceptable quality may so adversely affect our general HRS wheat quality that our *domestic and foreign markets will be lost to competitors* for many future years."[34]

The NDAES was not unmindful of the pressures on farmers to increase yields. The hybrid wheat program had a high-yielding, high-quality wheat as its goal, and for some time Station agronomists had been working with semidwarf genetic material with the same end in mind. That latter activity bore fruit in 1973 when the Station released its own semidwarf variety, which it called Olaf. Olaf marked a change in standard Station release practices. As Hazen noted, historically "the objective of this Station has been to name and release varieties which 'equal or exceed' previously named varieties in . . . wheat quality." Olaf was "inferior when judged by the quality standards used in . . . evaluation for high quality wheats." Still, its quality was acceptable and predictable, and it responded to farmers' yield demands. While the Station did not recommend wide planting of Olaf, it became popular quickly. In 1977 one in four hard red spring wheat acres in North Dakota was planted to Olaf, the popularity of which was exceeded only by Waldron.[35]

Olaf was essentially a compromise in which, as cereal technologist Orville Banasik notes, "a little less quality" was accepted to gain "a little higher yield." This was one of those cases in which the Station was caught between its desire to meet its clients' short-term requirements and its need to serve their long-term interests. The Olaf compromise did not settle the issue of yield versus quality. Cereal Technology chair Bert D'Appolonia notes that, while the "plant breeding team" and "the majority of producers are committed to quality," maintaining high standards in the face of economic realities is difficult. Parsimonious premium payments and yield-based government programs penalize producers of quality wheat while the rising merit of winter wheats and new baking technologies threaten to cancel their advantage. Whether the Station and the producers it serves can remain committed to quality remains an open question.[36]

The consuming struggle over wheat quality reminds us of the primacy of that crop in North Dakota agriculture and in the research efforts of the NDAES. But part of the story of the Hazen years was diversification. Since its earliest days the Station had encouraged diversification, and North Dakota farmers, always relatively innovative and progressive, had been open to the idea. The fifties, sixties, and seventies saw the introduction of a number of new crops in the state, most of which demanded an intensified research effort by the NDAES.

One of the major new crops was dry edible beans, such as northern, navy, kidney, and pinto beans. Dry edible beans had been produced sporadically and in small quantities in the area since at least the early 1940s. But they became something more than a curiosity crop largely as a result of the efforts and example of Max Campbell, an Oslo, Minnesota, farmer who began raising beans in 1960. In 1965 North Dakota farmers harvested 25,000 acres of beans, worth $957,000. Ten years later their crop from 122,000 acres was worth $17,390,000.[37]

Beans were attractive because they could be used in advantageous rotation with small grain crops, could be planted and harvested with small grain machinery, were easily stored, and found a ready and sometimes lucrative market. They also suffered from a variety of disease problems, however, and by the mid-sixties producers were demanding Station help. Eventually a full-fledged bean program was put in place, the centerpiece of which was an attempt to breed varieties that were "upright, high protein . . . , nitrogen-fixing, low flatulence, high yield, disease resistant or tolerant, adapted to a short growing season and non-discoloring in the case of pintos." Surely this was a tall order, with the nonflatulent bean especially seeming a scientific miracle to rival the space shuttle.[38]

Sunflowers were another crop that became important in North Dakota during Hazen's directorship. Sunflowers were attractive because they could be grown with small grain machinery, their seeds produced an attractive edible oil without cholesterol, and their byproducts had a variety of uses. Between 1962 and 1975 the production of sunflower seeds in North Dakota increased from 12,250,000 to 547,800,000 pounds. The earlier crop was worth $674,000 while the latter returned $59,102,000.[39]

Unfortunately, there were numerous obstacles to successful sunflower production, including "downy mildew, rust, sclerotinia stalk rot, and verticillium wilt .... the head moth, the midge, ... stem maggots and stalk borers .... bird damage, harvesting and handling problems." Producers and researchers have struggled with all of these problems, which have so far prevented the sunflower from reaching the potential many of its champions foresaw for it.[40]

Crop production was not the only area in which significant alterations occurred during Hazen's directorship. Major changes also took place in livestock production. Generally, North Dakota producers raised more beef cattle and fewer dairy cattle, sheep, and chickens. Between 1955 and 1975 the number of cattle being raised for beef on North Dakota acres increased from 1,350,000 to 2,511,000 while the number of cattle being milked plummeted from 410,000 to 122,000. Sheep numbers declined from 559,000 to 315,000, and the chicken population dropped from 3,979,000 to under 700,000. Swine numbers were generally stable during this period, fluctuating between 300,000 and 400,000 in any given year.[41]

To some degree these shifting numbers indicated changes in expectations and lifestyles among North Dakota farm families. Farmers were becoming more commercial and were less inclined to attempt to maximize family self-sufficiency. But market factors also played an important part in determining the livestock mix. Beef continued to rise in popularity among an increasingly affluent community of consumers. Sheep declined because of stiff competition between North Dakota wool and that of other producers and synthetics, and because a substantial American market could not be developed for mutton or lamb. Distances, low quality, and changing consumer tastes diminished dairy and poultry production in North Dakota.

Alterations in livestock numbers were reflected by changing research emphases at the NDAES. When Chair R. L. Bryant retired in 1966 Hazen abolished Poultry as an independent department and merged it with Animal Science. A similar fate befell the Dairy Department in 1970 when Chris Jensen retired. Dairy and poultry producers complained about the abolition of these traditional departments. Even though poultry and dairy products were less important overall, they remained significant to those who still produced them. Moreover, they might always bounce back in the future, as the small but healthy turkey industry seemed to forecast. Hazen acknowledged these arguments, but he concluded that resources were too scarce to be lavished on what had become marginal concerns in the overall scheme of North Dakota agriculture.[42]

Animal Science continued to be concerned mainly with the breeding, feeding, and marketing of the animals on which the NDAES still concentrated. Merle Light, for example, established an enviable national reputation with his sheep breeding program. The remoteness of North Dakota makes livestock marketing a major problem and shaves producers' profits. On the other hand, feed in North Dakota is relatively cheap. Bill Dinusson's work with feed supplements and with using pelletized barley as feed—for swine as well as cattle—provided North Dakota producers with insights that helped them overcome some of their locational disadvantages.[43]

Beef cattle producers had long believed that their enterprise was slighted at the Station relative to crop research. Part of the reason for what they saw as a misallocation of funds lay in the stubborn fact that crops were a more important factor in North Dakota agriculture than was livestock. But part of the problem lay with the producers themselves. As Bill Dinusson notes, "in the early fifties the cattlemen's association was kind of a friendship group" rather than an effective commodity organization. Moreover, "they always backed [the substation at] Dickinson, because of the Imperial Cass thing."* That began to change during the sixties when the cattlemen came under the aggressive leadership of Delbert Moore and Jack Dahl. The group demanded more attention to beef interests. They wanted a beef unit at the new irrigation research facility at Carrington and a new branch station devoted to grassland research. Hazen, who had just managed to get rid of a relatively unproductive branch station at Edgeley, was reluctant to create another. Branch stations could be useful research facilities, they made the work visible to clients, and they built political support in localities and among commodity groups, but Hazen wondered "whether this is the best alternative for use of funds and personnel resources." As he wrote Moore in 1974, "I have a continuing concern for the fact that we have not been able to appropriately and adequately support earlier commitments with adequate financing." Eventually the stockmen prevailed, however, and legislative appropriations in 1979 and 1981 allowed the creation of a branch station at Streeter devoted to grasslands research.[44]

In other areas research continued along set lines during Hazen's directorship. Irrigation continued to be a major research interest, with work centered at the new branch station at Carrington and at a demonstration site at Oakes, where experimentation was done with center pivot irrigation. Potato breeding, one of the great success stories of the Golden Age at the NDAES, continued to dominate

**The term "Imperial Cass" is a derogatory reference to Fargo, the seat of Cass County, reflecting the resentment many North Dakotans have historically felt toward that city for its relative wealth and political power. Because it was placed in Fargo the NDAES has been forced to bear the Imperial Cass stigma periodically.*

Horticulture, though traditional work with vegetables and ornamentals was not neglected. While the growing problem of salinization demanded increasing amounts of their attention, soil scientists concentrated on exploring the implications of nitrogen fertilization. Supported in part by chemical companies and other agribusiness entities, Station soil specialists conducted experiments that showed the potential of commercial nitrogen fertilizers for increasing yields and allowing the diminution of summer fallow. As with many advances in modern agriculture, this one had a dark side in the form of nitrogen poisoning of animals and contamination of water supplies, subjects that were also tackled by Station researchers.

The changing structure and economics of North Dakota agriculture provided important subjects for research by agricultural economists at the NDAES. Improved machinery allowed fewer people to do more work even while it increased farmers' credit needs and expenditures for off-farm inputs. Chemicals increased yields but also drove costs higher. Electrification improved rural living while further straining farmers' pocketbooks. Unit costs rose faster than prices much of the time, creating a classic "squeeze" and putting a premium on economies of scale, production efficiencies, and high yields. Farm sizes increased steadily as thousands left agriculture. The days when one could maintain a modern standard of living and make efficient use of capital, labor, and machinery inputs on a half, or three-quarter section farm in North Dakota were fast passing away. Perhaps the most remarkable facet of the exodus from farming was that it did not lead to decreasing land prices. Every time a farmer sold a farm, three or four neighbors saw acquisition of it as a means to greater efficiency. The resulting competition drove prices up. In 1961 real estate prices were "*9.7 times net income per acre; in 1954-55, the ratio was 7.7, and for 1949-50 it was 5.5.*" That trend continued and intensified over the next twenty years.[45]

As agricultural production became a larger, more complex, and more heavily capitalized business, North Dakotans wondered whether the family farm had a future. Agricultural economist Fred Taylor thought it did, arguing in 1960 that "these trends are strengthening family farming in North Dakota, to the extent that farms of efficient size and more adequate incomes . . . are created . . . . Present size farms offer more adequate returns on capital and labor and provide a better living for the farm families that . . . operate them." The question was, how far would the tendency toward fewer and bigger go? The farm that was adequate yesterday was inadequate today; one that provided a reasonable living today probably would not tomorrow.[46]

Agricultural economists generally avoided policy questions after the mid-fifties, suggesting instead how farmers could become more cost efficient producers and marketers. Thor Hertsgaard, for example, studied marketing patterns for poultry and swine, and Fred Taylor and an associate discussed how improved highway transportation and the St. Lawrence Seaway had changed the marketing of North Dakota grain. Rainer Schickele and his colleagues had been at least a bit nostaligic and had worried about the trends they saw in agriculture. Rather than fearing the impact of new developments on farming, Station economists during the Hazen years saw in these new opportunities for farmers. As early as 1960, for example, future College president Laurel Loftsgard saw the computer as a potential aid in farm planning.[47]

Station agricultural economists began to move beyond farming as such, investigating a range of economic development issues. A 1957 study of the potential for a straw pulping industry was the first in an impressive series of investigations of the feasibility of developing agriculturally oriented industries in North Dakota. Station researchers also got involved in the effort to develop tourism in North Dakota, concluding in 1969 that only unpleasant weather and a lack of scenery prevented the state from becoming a thriving vacationland! During the seventies Station social scientists and natural scientists joined forces in exploring the impact of coal and oil development in western North Dakota.[48]

The dedicated, multifaceted, and productive work of Station scientists represented what Hazen liked to call a "handsome dividend" on the state's investment in the future. To his mind, public funds devoted to agricultural research were less an expenditure than an investment because they were returned to the state many times over. Hazen took a good deal of satisfaction from the strong "moral support" the state provided the Station, as well as from impressive increases in funding that allowed the staff to be expanded and facilities to be upgraded. Between 1960 and 1975, he informed one correspondent, "the number of scientist man years supported by this Station had increased from 74 to 120."[49]

Still, Hazen was frustrated with some aspects of the research program. The second half of his tenure was a time of severe inflation. Appropriations rose—sometimes apparently dramatically—but costs usually rose at least as fast. Answering an inquiry about funding in 1977, Hazen complained that "available funds have not kept up with the cost of inflation and other costs of maintaining a research program. Most notable are lack of authority and funds to pay competitive salaries and adequate facilities and equipment for scientists to utilize." The salary problem, a perennial one in North Dakota, was especially vexing because successful Station performance depended on quality personnel. "The importance of experience and competency of . . . the highly trained professional personnel, cannot be overemphasized in a continuing agricultural research program," he wrote state budget director Lloyd Omdahl in 1976. "Excessive changing of competent personnel can be devastating to the mission and success of such programs."[50]

The federal government was of little help. Generally, Washington demanded more, gave the stations less discretion in spending, and provided fewer real dollars. In 1973, when the Nixon administration proposed a 20 percent cut in federal funds for station research, Ken Gilles fumed that the Station was "astounded at the degree of inconsistency shown by the Executive Branch . . .when, on the one hand, [it] encourages rural development, encourages increase in agricultural exports . . . and, at the same time, . . . implements a plan to reduce support for agricultural research." Even when money was appropriated it was not always spent. When Nixon's Office of Management and Budget impounded weed research funds in 1973 Senator Young commiserated: "Arlon, the Administration has cut many valuable programs and seems to pick out the best programs and those areas where it hurts most." Unfortunately, even the powerful Mr. Wheat had to admit his inability to pry the money loose.[51]

Hazen also complained that research productivity did not advance in proportion to the funding increases the Station received. A man of high integrity and a well-developed sense of duty, Hazen wanted to see a full return on every dollar invested. To some extent he viewed the problem in rather a cranky, old-fashioned way, complaining of "shorter work hours, more holidays [Station personnel had stopped working Saturdays in the late fifties], [and] greater individual 'rights and benefits.'" But he also recognized that the easy questions had been answered and that future accomplishments would not come as quickly. As he remarked to one correspondent, "the task of the agricultural research scientist has become increasingly far more complex to arrive at usable and applicable research results."[52]

Even in these relatively productive and satisfying years, then, there were difficulties and frustrations for Hazen and the Station. As the sixties proceeded and turned into the seventies the source of frustration for Hazen and his staff came increasingly to be located outside of North Dakota.

III

During the forties and fifties the publicly sponsored agricultural research system was remarkably free of criticism. Science was generally held in high regard—sometimes in awe—and few questioned what agricultural researchers were doing or how they were doing it. Scientists were heroes during the war and the postwar era. The public was satisfied with them, and they were satisfied with themselves.

All of this began to change in the turbulent sixties, and by the early seventies the agricultural research system found itself besieged by enemies without and beset by doubters within. Part of the problem could be attributed to the trend of the times. This was a period when every institution in American life came under scrutiny and often criticism. But it was also the case that the critics raised substantial issues—albeit sometimes in a shrill and poorly reasoned manner—that needed to be addressed.[53]

One set of critics focused on the failure of the agricultural research system to protect the environment sufficiently, ensure food safety, or attend adequately to nutritional considerations. What might be called the "consumer protest" against the system can be traced to 1962, when Rachel Carson published her enormously influential indictment of DDT, *Silent Spring*. Carson's book was the major early stimulant of what became an enthusiastic and popular crusade against environmental degradation. Joining the environmentalists were those concerned about the impact of chemicals in food on humans who consumed it and those who wondered whether research aimed mainly at increasing production provided food of optimum nutritional value. By the mid-seventies other critics were wondering whether, in light of the energy crisis, an agriculture dependent on petroleum-based fuels and chemicals could even be sustained. These types of criticism indicated a basic shift in popular American attitudes regarding science. What had once been seen as a benevolent force serving humanity was now perceived as a dangerous to life. As scientific institutions that had been deeply involved in the promotion of chemical agriculture the experiment stations obviously came under heavy criticism.[54]

Another set of outside critics raised questions about the social consequences of agricultural research. In 1973 Jim Hightower of the Agribusiness Accountability Project published *Hard Tomatoes, Hard Times*. Focusing on the research of the California Agricultural Experiment Station, Hightower contended that the "land grant complex has come to serve an elite of private, corporate interests in rural America, while ignoring those who have the most urgent needs." More specifically, Hightower indicted the system for directing its research to the needs of food processors, large farmers, and agribusiness corporations, to the detriment of small landowners, tenants, farmworkers, and consumers. The sixties and early seventies was a time of sympathy for the underdog in the United States, and *Hard Tomatoes, Hard Times* showed that such sympathy was being extended to agriculture.[55]

External critics were soon joined by internal doubters. There were those in the system who worried about environmental degradation and the social consequences of research, and some of them were emboldened to speak out. More troubling were critics within the scientific community who questioned the quality of science in the USDA and the experiment stations. The short-lived food crisis of the early seventies led to a flurry of finger pointing and assessment of blame. Panels of elite scientists found the agricultural research system culpable and damned it for "pedestrian" investigations, devotion of insufficient funds to basic science, "duplication of effort in some areas and . . . a wastage of resources." Everybody seemed to agree that the agricultural research system, effectively above criticism just a few years before, was seriously deficient. They just could not agree on what the major deficiencies were.[56]

Mark Twain once said that he hoped to be in Cincinnati on the day the world ended, because everything happened ten years later there. The same might be said of North Dakota. The sharp conflicts and the shrill criticisms of the sixties and seventies were only faintly echoed in the Flickertail State, though some of them have endured and gathered strength in the eighties. The NDAES was largely immune to the sorts of criticisms other state stations were receiving. North Dakota was a producer state in which consumer protection was not an especially cogent or popular issue. Few could believe that environmental degradation was a problem in a land of deep blue skies and wide open spaces, and organized environmentalists were a negligible force. The structure and social configuration of North Dakota agriculture was so divergent from that of California that Hightower's remarks had little relevance. Engaged in production agriculture and seeking quick solutions to immediate problems, North Dakota farmers were not disturbed by charges that experiment station research was pedestrian and overly applied. Indeed, those very qualities were the ones most clients wanted to see emphasized. But the Station was also part of a national enterprise that was under fire. It felt the criticisms of the time, was stung by them, and tried to respond.

The varied pressures for change in the research system increased through the sixties and seventies. Neither Arlon Hazen nor most of his subordinates had much appreciation for critics' points of view. Like most station directors around the country, Hazen was unused to stringent criticism of the system. He was, moreover, a conservative, production-oriented administrator who believed his institution was doing the right things well. With some justification, he saw criticism as a passing fad, writing one farmer that among "writers, politicians, planners, environmentalists, and ecologists . . . . To thump the food and fiber producers is a kind of literary and oratory [*sic*] sport."[57]

The NDAES was devoted to serving the farmers of North Dakota. It did so mainly by dealing with the production problems that beset them all to a greater or lesser degree. The Station had learned the painful lessons of its stormy past. It conceived its role narrowly and carried it forth carefully. It took care to alienate no powerful individual or group and was always conscious of the importance of maintaining credibility with North Dakota farmers. To Hazen this meant listening to all and working with all while showing special consideration and favors to none. He eschewed potentially compromising relationships, especially with agribusinesses, refusing "proprietary research" and barring Station scientists from positions "as paid members of boards of directors of agribusiness firms." Its very responsiveness to the public and its anxiety to maintain credibility eventually forced the Station to respond to critics. Before that happened, though, the federal government responded, and that had an impact on the Station.[58]

As early as the 1960s the federal government began implementing policies designed to improve research performance throughout the system. These efforts reached a climax in the sweeping, multifaceted Food and Agriculture Act of 1977. One of the goals of this legislation was to induce greater planning and coordination—and thus less duplication—in the system by creating new bodies to oversee public agricultural research. Few state experiment stations were enthusiastic about this goal or the means of achieving it. Planning and coordination sounded good, but many directors believed that the existing system had worked well and that its apparently chaotic structure actually made it highly flexible. When people at the NDAES heard words such as "planning" and "coordination" they thought about the Great Plains Agricultural Council, an entity created in 1946 that was supposed to bring unity to agricultural research efforts in the region but that proved productive mainly of boring meetings and pointless paperwork. They were understandably cynical. The fact was that dramatic improvements in planning and coordination were very difficult to achieve voluntarily, and that was the case in the sixties and seventies as it had been in the thirties and forties. Indeed, attempts at better planning during the sixties and seventies had the ironic effect of disrupting the system and making it more chaotic. When the Carter administration's planning efforts reached a hitherto unattained crescendo of confusion in 1980 the powerful North Central Region Association of directors decided to cease cooperative planning efforts altogether.[59]

Another means of improving research performance in agriculture to which federal policymakers were drawn was the replacement of at least some formula funding with competitive grants. This was another idea advanced markedly by the Food and Agriculture Act of 1977. The elite in the scientific community believed that formula funding—i.e., the provision of federal monies in stable amounts on a year-to-year basis under the provisions of legislative enactments such as the Hatch Act—discouraged innovative, cutting-edge research. Wide competition for funds would presumably force scientific institutions to move beyond the pedestrian, largely applied agricultural research in which their critics believed they were mired.

The experiment stations argued that formula funding provided the fiscal stability necessary for long-running research programs. Many station scientists feared these programs would be jeopardized if more money went into competitive grants. They also believed such grants would go mainly to elite institutions with strong support staffs, that they would make science more competitive and less cooperative, and that they would encourage unrealistic, miracle-a-day expectations in the public. As the director of an underfunded, nonelite institution with long-term research commitments, Hazen was understandably concerned about the threat to formula funding. As early as 1966 he bemoaned Washington's "attitude of indifference to the traditional concept of payment to the states, apparently based upon the proposition that this kind of use of federal funds would not create and build excellence." His concerns about the future of formula funding only intensified with the passage of time.[60]

Hazen was also disappointed with the growing tendency of Congress to earmark portions of formula funds for particular social purposes or to place controls on how money was spent. For example, he saw the congressional emphasis of the sixties on rural poverty work as irrelevant to North Dakota and he carried on a running argument with the USDA regarding how rural development funds could be spent. Earmarking funds and attaching strings to them was another tendency that reached a climax with the Food and Agriculture Act of 1977. As one perceptive scholar has remarked, that piece of legislation "included a little of everything for everyone."[61]

Changes at the USDA indicated the responsiveness of that agency and

Congress to public criticism. Other areas of the government also reflected the shift in public attitudes. Consumer and environmental protection became increasingly important concerns, resulting in the creation of the Environmental Protection Agency in 1971. To the scientists at the NDAES the actions of the EPA seemed to operate mainly to take away the weapons with which insects and weeds were fought. Associate Director H. Roald Lund complained strenuously of EPA plans to ban DDT and 2,4,S-T in 1971 and weed scientist John Nalewaja bemoaned the fact that the EPA never asked for Station input. Hazen believed that the EPA had a short-range view of the consumer's interest that ignored the necessity of maintaining food production. Obviously, these are issues that have not gone away, and entomologist Ted Schultz is but one researcher who worries about the implications of a "philosophy that we should live in a risk-free environment."[62]

Even as Station scientists complained of the new social concerns and their implications, they responded to them. Their apparent reluctance to do so earlier derived from the stridency of critics of the system, the apparent satisfaction of North Dakotans with the Station, and the fact that resources were limited. One very positive aspect of the new social concerns and demands from an institutional standpoint was that they resulted in some new funding. As always, the Station proved willing to expand the scope of its work when needs appeared and resources became available. By the mid-seventies NDAES researchers were working in such new areas as wetlands management, water quality, reclamation of strip-mined lands, biological pest control, development of alternative energy sources, and no-till agriculture.[63]

Environmental consciousness at the Station grew impressively during the sixties and seventies. If nothing else, Rachel Carson's attack on DDT reminded NDAES scientists of the health hazards of chemicals. Safety training began for Station and Extension personnel in 1964, forming the basis for an ambitious program of public pesticide safety seminars for farmers. In 1970 and 1971 the Station demonstrated its ability to carry forth impressive environmental research when it undertook a study of mercury in North Dakota soils, crops, and livestock tissues. Early in 1971 the Station devoted an entire issue of *North Dakota Farm Research* to the "three 'E' terms, Earth Science, Ecology, and Environment." This issue drew criticism from an unexpected source, when right-to-life activists complained of sections favoring population control.[64]

The NDAES also demonstrated sensitivity to consumer concerns. Much of what might be called consumer-oriented research was carried out by members of the College of Home Economics. The history of home economics research at the Station is a checkered one. In 1925 the Purnell Act provided federal funds for home economics research, prompting the Station to support food and nutritional investigations. During the depression the NDAES was given permission to terminate home economics research without losing funds on the grounds of financial exigency; such research was essentially moribund at the Station for a generation thereafter, and Hazen was not inclined to change the situation. Cautious and conservative by nature, he was reluctant to expand research commitments and he was unimpressed by the abilities of some of the researchers in home economics. Former associate director John Callenbach recalls that Hazen believed "they didn't have the talent to do the work—the proposals they sent in were rubbish."[65]

That situation began to change in the 1960s as a result of a number of factors. The increasing consumer consciousness in society, reflected eventually in the provision by Congress in the Food and Agriculture Act of 1977 of new funds for home economics research, was a major consideration. Another was the impressive upgrading of the home economics staff. A third was the formidable presence of Home Economics dean Katherine Burgum, who realized what her researchers could do and who knew how to help them do it. Burgum was a firm and articulate advocate of her college, whose considerable political skills had been honed during her years of service in the upper echelons of the North Dakota Republican party. Her personal attributes served her college well. As Callenbach puts it, "Katherine Burgum got funds through her political skill and bulldog tenacity."[66]

The work undertaken by Station researchers from the College of Home Economics has been of three basic types. Researchers in Textiles and Clothing, Food and Nutrition, and Family Economics have done what might be broadly styled as consumer-oriented investigations. Other researchers, especially in Food and Nutrition, have done a good deal of work in food processing and marketing, areas in which general Station interest has become more pronounced. Most of the social science research sponsored by the Station has been carried forward by the Department of Child Development and Family Science, which has concentrated especially on family dynamics.

Station research in home economics increased dramatically, with funding rising by 150 percent between 1979 and 1985. At that point, Dean Charlotte Bennett likened the college to "a small jewel that is beginning to sparkle a lot." Substantial budget cuts by the 1987 legislature were felt with especial severity by home economics researchers in the Station, who were victimized by the LIFO principle (like many institutions, the NDAES generally cuts program expenditures on a "last in, first out" basis) and perhaps as well by the fact that they did not have strong relationships with powerful commodity groups. Still, home economics research at the NDAES is likely never to return to the figurative wilderness in which it found itself before 1970, because it simply has too much to offer North Dakota's economy and society to be ignored.[67]

IV

Arlon Hazen resigned the directorship of the NDAES on November 1,

1978, to assume the position of director of the North Central Region Association of Experiment Station Directors. That appointment indicated the high regard in which his peers held him. A few weeks after his resignation he was felled by a fatal heart attack.

Before he died Arlon Hazen had the opportunity to tell what his service to North Dakota had meant to him in a special issue of *North Dakota Farm Research.* In that same issue his chairs and superintendents had the opportunity to tell him what he had meant to them. They held him in high regard, and rightfully so. He was a man of loyalty and integrity, and all who knew him respected those qualities. Hazen built on the foundations laid by his predecessors, creating the sort of scientific institution of which they would have been proud. In that sense he was a steward who took a going concern and made it better. He was a prudent and careful director. He emphasized service to the state and favored the sort of research he thought would best serve the state—research to help farmers and ranchers produce and market more and better crops and animals and thus achieve greater income security. The Station under his leadership offered what most farmers wanted. They trusted it, they used it as never before, and they supported it before the legislature.

In a sense Hazen was a director for his times. His tenure witnessed the rise of commodity groups, increasing commercialization in agriculture, and incessant calls for higher yields. Practical times demand practical people, and he was that. But he also knew when to be flexible—when to change and strike out in new directions. Like his two greatest predecessors, John Henry Worst and H. L. Walster, Hazen was a man of strong opinions. But his strongest was that the Station must serve the people and that they ultimately must judge how they want to be served. In that sense Hazen was a director for all times.

Arlon Hazen, 1978.

Agricultural economist and NDSU president L.D. Loftsgard, about 1985.

Agronomy technician Olaf Gronaas (l) with wheat breeder Glenn Smith (r) in 1962. The semi-dwarf hard red spring variety "Olaf" was named for Gronaas.

Agronomist Earl Foster, 1980.

Agronomist Jack Carter (l) and range scientist Warren Whitman (r), 1961.

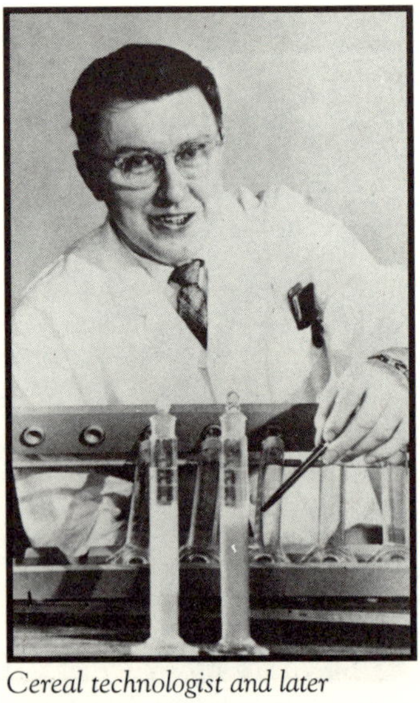

Cereal technologist and later NDSU vice-president for agriculture Ken Gilles, 1962.

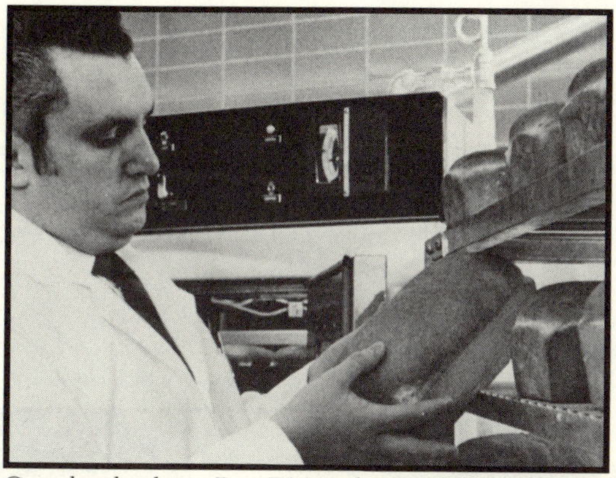

Cereal technologist Bert D'Appolonia, 1970.

Biochemist Harold Klosterman, about 1960.

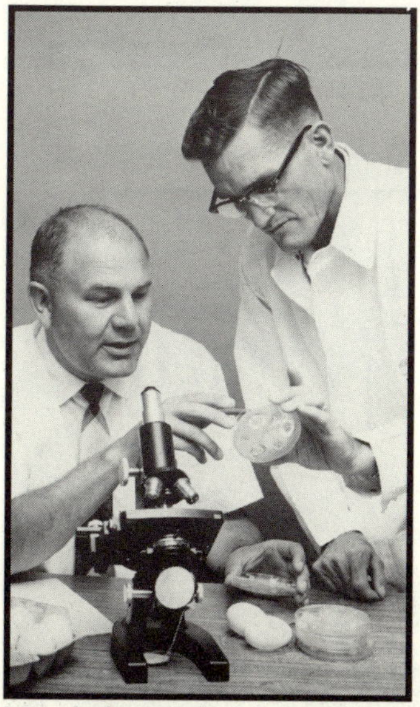

Microbiologist I.A. Schipper (l) and poultry husbandman W.C. Lockhart (r) dicussing a problem in 1962.

Horticulturist Neal Holland uncovering some impressive squash in 1972.

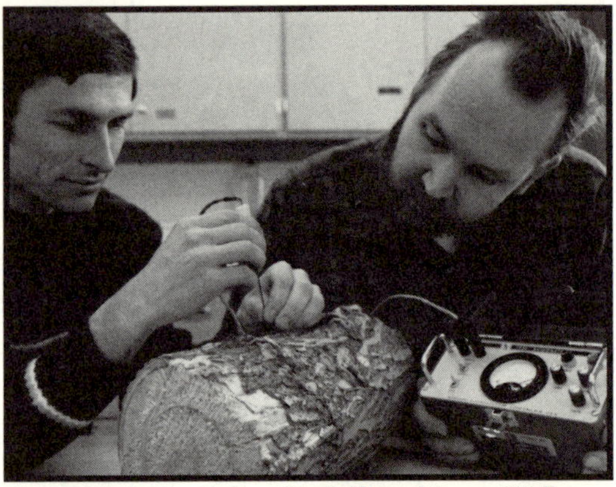

Plant pathologists Robert Stack (l) and James Walla (r), tracking down a culprit in 1977.

Sheep specialist Merle Light with the Station flock, about 1970.

Soil scientist Enoch Norum, 1975.

Agricultural economist Fred Taylor, about 1980.

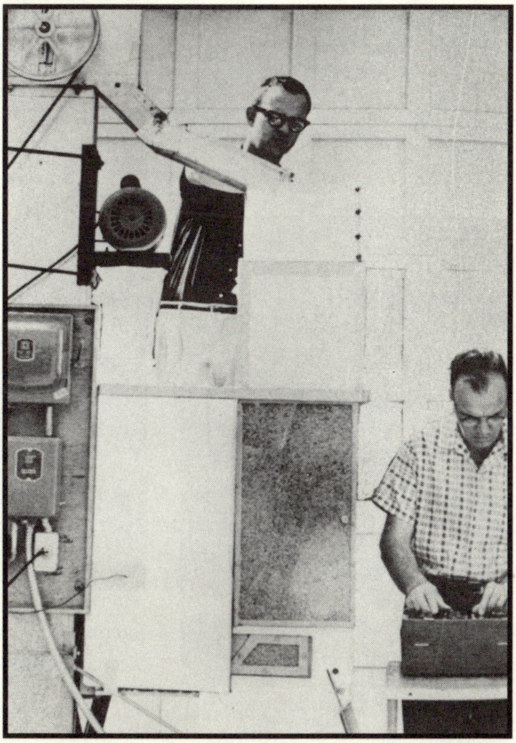

Agricultural engineers George Pratt (above) and Bill Promersberger(below) in 1959.

Animal scientist Clayton Haugse entering the computer age, 1970.

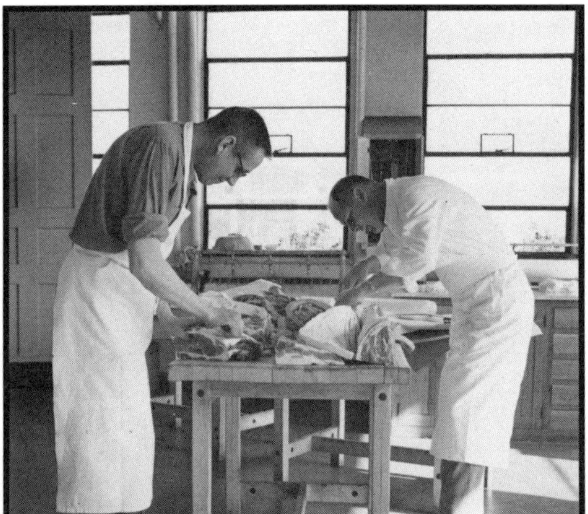

Animal scientists Verlin K. Johnson (l) and John N. Johnson (r) using more traditional tools of the trade in the late sixties.

Potato breeder Robert Johansen (l) inspecting one of his new varieties with research technologist Bryce Farnsworth, about 1985.

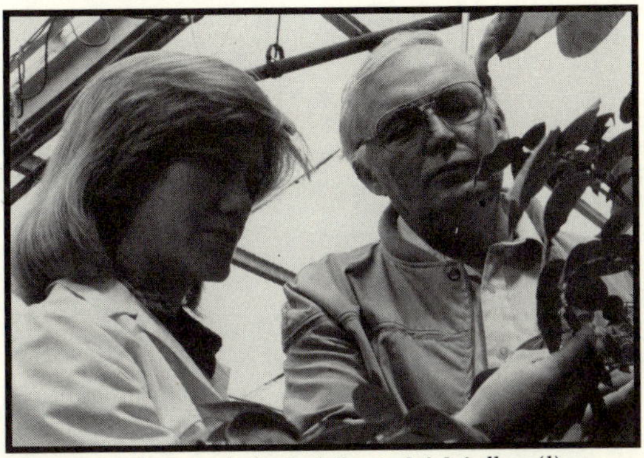

Extension plant pathologist Marcia McMullen (l) consulting with plant pathologist Richard Kiesling (r) on a foliar problem, about 1985.

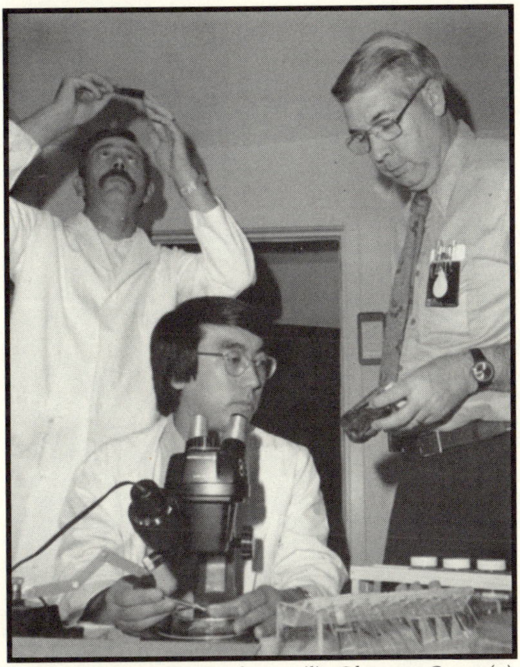

Entomologists A.W. Anderson (l), Christian Oseto (c), and Greg Mulkern (r), 1976.

Weed scientist John Nalewaja in a wild mustard field, 1972.

9

TOWARD THE FUTURE

All of the original dollars come from the sun and the soil. The new dollars are basic to the economy. There's no research more important than agricultural research. The North Dakota Agricultural Experiment Station is the single most important fountainhead of wealth in the state.

William Guy, 1987

The early people here were committed to serving the public, and that has never diminished....All of the technical expertise in the world can't compensate for a commitment to service.

H. Roald Lund, 1987

H. Roald Lund succeeded Arlon Hazen as director of the North Dakota Agricultural Experiment Station. Lund is the first director to be born in North Dakota and to receive most of his training at the NDAC. Born in 1933 on a farm near Hillsboro, Lund took bachelor's and master's degrees at the NDAC and held a research appointment at the Station before going on to Purdue for a Ph.D. in agronomy. In 1965 he returned to take a regular appointment in the Agronomy Department, where he specialized in corn. Lund became associate director in January of 1970 at Hazen's request. He was named acting director when Hazen retired and received the permanent appointment a few months later.

Lund presents a noteworthy contrast to Hazen in terms of personality. He is an extroverted, gregarious person who genuinely enjoys talking to people, either individually or in groups. His aggressively friendly style and infectious sense of humor present a striking contrast to the more phlegmatic and diffident character most NDAES researchers and most North Dakotans manifest. Lund is an evangelist for the Station and for North Dakota agriculture, and he sees public relations as a major part of his job. Shortly after he became director he initiated the practice of traveling extensively throughout the state, speaking to public meetings about the work and the needs of the Station. "His big job has been stumping the state . . . ," one Station veteran notes; "Hazen wouldn't work the state like that."[1]

Differences in style between Lund and Hazen are misleading, however, because they are very similar on matters of substance. Lund is as committed as his predecessor was to research in production agriculture. He shares Hazen's devotion to service to North Dakota as well as his recognition that serving the state is the Station's primary job. When Lund says that "all of the technical expertise in the world can't compensate for a commitment to service," he is subscribing to an article of faith promulgated in the first years of the Station and kept scrupulously by his predecessor. Lund also shares Hazen's recognition of the political responsibilities of the director. He knows that while a director should be a scientist, he must be a politician. Hence, Lund has been attentive to major farm and nonfarm groups. He has worked closely with elected and appointed public officials and has gained the respect of many of them. State senator Bryce Streibel, for example, calls Lund a "tremendous communicator" who is "polished and organized" and who has a remarkable "ability to perceive the whole." Lund has also worked effectively for the NDAES and for North Dakota in Washington. His close working relationship with Senator Quentin Burdick, for example, helped bring funds for the Northern Crops Institute, durum and feed mills, and weed and sunflower research facilities to the campus.[2]

Within the Station Lund sees his role as that "of a coordinator and facilitator" rather than "an authoritarian director." He searches for "consensus" and seeks "to make it easier for people to respond to problems." Like Hazen, he delegates responsibility liberally to department chairs and to his assistant deans, but he remains the key figure in setting directions. The enhanced commitment to basic research at the NDAES, for example, reflects Lund's perception of where the Station must go and represents a major change of direction for the institution. Clients generally find Lund responsive and respect him. "He's accessible . . . ," Agricultural Stabilization and Conservation Service Director and prominent Northwood farmer Keith Bjerke says; "He'll do something about your complaints if he can. He's very direct, and he doesn't always tell you what you want to hear."[3]

When Lund was named permanent director he promised "to maintain an optimistic outlook, a positive philosophy, and . . . to continue to try to respond to the directives as determined by you, the consumer of new information." Keeping that promise has not been easy, because in retrospect it may well be clear that Roald Lund served as director during some of North Dakota agriculture's most trying times as well as during a period when the institution he headed underwent fundamental and difficult changes.[4]

North Dakota agriculture went from the penthouse to the outhouse during the late seventies and the eighties. When Lund became director agriculture enjoyed rare prosperity. Yields were high and prices for most crops were good, if not spectacular. Double-digit inflation kept interest rates high, but land values rose even faster. By 1981 the average North Dakota acre was worth $454, and in the southern Red River Valley land values rose to the unprecedented level of $1,125 per acre. People who had what they considered modest operations became millionaires—on paper. Few predicted that the boom would bust. Former Secretary of Agriculture Earl Butz was only the foremost of a number of apostles of expansion who told farmers to "get bigger, get better, or get out." Lenders urged farmers to expand, sometimes threatening to drop those who would not buy more land or buy as much more as the lender thought they should have. Farmers were urged to borrow heavily on their equity to purchase more land. It did not really matter, they were told, that a high-priced acre could not produce enough even to pay the interest due on it. Galloping inflation and a hungry world would solve this minor problem soon enough. North Dakota farmers are cautious. Their history has given them a healthy respect for economic and environmental vagaries. But some expanded—too much, ultimately—convinced that the future belonged to the big and that their children's futures in agriculture depended on them enlarging their operations in the present.[5]

The situation began to change in the late seventies. By the early eighties North Dakota agriculture was suffering as it had not suffered since the terrible years of drought and depression half a century before. The reasons for the agricultural collapse and the ensuing depression are complex. Part of the problem was rising world food supplies and a subsequent softening of prices. Then there were the Ford and Carter embargoes on grain sales to the Soviet Union, which raised doubts about the reliability of the United States as a supplier. In 1979 the Federal Reserve Board decided to fight inflation with high interest rates, a policy that succeeded ultimately but at the cost of slowing the economy. High interest rates and the heavy borrowing to finance budget deficits initiated by the Reagan administration resulted in a rapid repatriation of dollars and a dramatic rise in the relative value of American currency. The overvalued dollar suffocated American exports, among which agricultural products are prominent.

The effects of these complicated developments on agriculture in North Dakota and elsewhere were catastrophic. Prices plunged. Interest rates stayed high but inflation—the debtor's strongest ally—dropped dramatically. That extravagent hopes were turning to gloom was reflected in plummeting land prices. In one year alone, Station economist Jerome Johnson reported, the value of land in the southern Red River Valley declined by 16 percent. Between 1981 and 1985 the value of the average North Dakota acre fell by 26.2 percent. The situation was especially serious for those who had borrowed to expand. Fixed obligations could not easily be met with prices declining.The value of purchased land and of equity used for collateral shrank steadily. Individual visions of future strength and security in agriculture melted away, replaced by doubts about present survival that were all too real.[6]

The Reagan administration, committed to a free market for agriculture, ironically dealt with the farm crisis by throwing money at it—more money ultimately than had been spent on farm programs by all previous administrations combined. Producers increasingly "farmed the government," but that brought survival rather than prosperity. For some it did not even bring survival. In North Dakota alone hundreds were forced out of agriculture. Disproportionately represented among the victims of the farm crisis were young, educated farmers and forward-looking managers who had followed the advice of Butz and others in the know. In this situation the optimism of Roald Lund and other enthusiasts for North Dakota agriculture seemed to many to be unrealistic or inappropriate.

During this time of agricultural deterioration some observers began to question the emphasis at the NDAES on production agriculture. Overflowing bins of unmarketable crops made farmers wonder why the Station continued to search for innovations to allow greater production. The Station recognized that production problems would always exist and that production research must therefore continue. But the NDAES also responded to the concerns of its constituents, undertaking or more heavily emphasizing research on marketing, alternative uses for agricultural products, economic development, and control of production costs.

This shift was not as dramatic as it might initially appear. The NDAES had always emphasized improving incomes and enhancing income security for people in an environment and an occupation heavily laden with risk. Addressing production problems was an important step toward achieving those goals and so, too, was research in marketing, alternative uses for production, diminution of costs, and so forth. Former governor Bill Guy contends that the Station "is the single most important fountainhead of wealth in the state." That position carries a responsibility the NDAES has always taken seriously and has fulfilled in a number of ways.[7]

Within the realm of production research the eighties saw important changes. The NDAES placed unprecedented emphasis on basic science. Station researchers began work in the new, exciting, and risky regions of biotechnology. Mirroring the shift in research emphasis was a changing of the guard, as the generation that had dominated the institution during the fifties, sixties, and seventies began to retire from the scene.

The farm crisis affected more than the direction of research. For a century the funding of agricultural science in North Dakota has waxed and waned with the fortunes of the farm economy. Until the mid-eighties state appropriations for the NDAES held up remarkably well, but then cuts came. Budget reductions forced some difficult and painful curtailments of programs and led the Station to contemplate means of establishing a more secure financial base for the future.

As the NDAES looks to that future it sees uncertainty. New and unproven research emphases and shaky funding are not the only causes for doubt. As a nation, the United States seems decreasingly committed to agricultural research and the values that have sustained it. Dwindling numbers of farmers and the continuing shift in the American economy away from agriculture add further concerns.

But an uncertain future is nothing new for the NDAES. It embarked on a journey over uncharted waters in 1890 and suffered more than its share of hazards, but it always followed the guiding star of service. It follows the same star as it sails into the uncharted seas ahead.

I

The agricultural depression revived the debate of fifty years before regarding the appropriateness of production research in the system, and the NDAES has not escaped criticism. "If there's one area where I would be somewhat critical, it's in this area of continually emphasizing greater yields," notes former Governor Art Link. "Sheer volume is not the answer." James Brokaw, a legislator and cattle raiser from Forbes, complains that "they do everything under the sun to get you to grow more . . . . They want us to raise the supply but they don't do anything about demand. When people lose their farms, that's not progress." Cogswell farmer and state representative Scott Stofferahn argues that production research focuses on the individual producer and ignores the structural aspect of the farm problem. "They don't want to admit that the problem is structural, not individual," Stofferahn contends. Even some Station veterans bemoan the stress on production research. "We emphasize production too much," Charles Edgerly believes. "The way our production has increased, it's not surprising farmers are going out of business. We have to pay attention to the net return."[8]

The NDAES was sensitive to criticism of its research emphasis. In the summer of 1984 an entire issue of *North Dakota Farm Research* was devoted to the economic crisis in agriculture. This was certainly after the fact, but it was a quicker acknowledgment of the problem than most experiment stations showed. Agricultural economists suggested management strategies for dealing with the crisis. While suggestions of management changes did not satisfy those who focused on major structural problems in agriculture, it should be noted in fairness that an individual experiment station—let alone an individual farmer—could do little about national or international economic arrangements. In response to the crisis, Station researchers also gave thought to such problems as the growing necessity of off-farm employment and family stress.[9]

Other areas received increased attention as well. Reminding the state of the economic vulnerability that accompanied its dependence on agriculture, the farm crisis stimulated a greater interest in economic development. The NDAES was heavily involved in that effort and intends to remain so. In addition, renewed emphasis was placed on finding alternative uses for agricultural products. The Station was involved particularly in studying fuel alternatives to petroleum that could be derived from North Dakota products. This research, which continues, has focused especially on ethanol and sunflower oil. Director Lund foresees that the search for alternative uses for North Dakota crops will become an increasingly important Station concern over the next few years.[10]

The agricultural crisis also led to a heightening of traditional Station concerns. Interest in marketing problems, for example, once more came into prominence. A major step forward in marketing was taken in 1981 when the Northern Crops Institute was established and facilities for it were constructed on the NDSU campus. The NCI is a cooperative venture involving the states of North Dakota, Minnesota, Montana, and South Dakota, as well as a number of commodity groups and general farm organizations. The purpose of the institute is to promote such "northern crops" as durum and hard red spring wheat, flaxseed, barley, edible beans, and sunflowers among potential customers. The NCI boasts attractive reception facilities to acquaint potential foreign and domestic buyers with regional crops, as well as laboratories to demonstrate processing and the adoption of local products to customers' needs. The NDAES has been heavily involved with the NCI since the beginning, with cereal scientists, agronomists, and agricultural economists being especially active.

Another area of traditional interest that once again received emphasis in Station research was that of profitability. A perusal of Station literature will convince the reader that the NDAES has never been unconscious of the issue of net cost and net profit. What the downturn of the eighties did was to increase concern with these issues. Profitability and marketing promise to be major themes in Station research well into the future.

These shifts in emphasis represented a significant Station response to clients' concerns. They did not signal a fundamental redirection away from production research, however. In spite of the agricultural depression production research has remained the primary thrust at the NDAES. There are several reasons why that is the case.

For one thing, the stubborn truth is that the NDAES has limited flexibility. Research programs are fixed, grants from agribusiness entities and commodity groups are in place, scientists have tenure, and funding does not increase dramatically. Those realities limit the ability of the Station to respond to new problems unrelated to its traditional emphases. Inflexibility is a fact of life in most large institutions. Thoughtful observers of the NDAES such as former agriculture commissioner Myron Just worry that inflexibility will lead to ossification, and that very real danger is one of which the Station must always be conscious. Being a public institution dedicated to service helps the NDAES in this regard.[11]

Were the Station to model itself more closely after the business corporation it might gain greater flexibility. If tenure did not exist and if longstanding programs could be terminated quickly it stands to reason that Station response time could be cut. One problem with this approach would be that the termination of programs temporarily out of fashion would entail a waste of invested resources. Another problem would be that it would make the NDAES a much more stressful place to work, and that would have implications for the state. It is unlikely that researchers would develop strong loyalty to an institution that showed little loyalty to them. They would likely pursue other opportunities more vigorously and would be disinclined to serve the state in the multitude of immeasurable but significant ways they serve it now. It is also reasonable to assume that scientists working in an insecure, high-risk public environment would demand similar compensation to that received by their counterparts in private industry. Whether the state could afford such an establishment is doubtful.

A dramatic shift away from production research would result in a Station that looked very different. The departments that have been at the Station's core—Agronomy, Animal Science, Entomology, and Plant Pathology, for example—would be required to slip into obscurity or perhaps fade away altogether. All of the training and thinking of most agricultural researchers is toward solving problems that impede production and it is difficult to imagine a fundamental shift in that habit of mind. It is as difficult for the average agricultural scientist to stop thinking about production problems as it would be for the average minister to stop thinking about saving souls or the average teacher to stop thinking about student learning. It is at the core of what they do. "You can't expect experiment station scientists to be all things to all men," Tom Conlon argues. "Others are responsible for marketing and selling. We're supposed to improve agriculture, and you don't do that by being less productive."[12]

It is also the case that many observers of North Dakota agriculture and of the NDAES hold that production research is appropriate and even necessary. Some perceive an irony in the fact that we criticize production research not because it has failed, but because it has succeeded so well. The high yields that have represented that success have led us to take production research for granted. As Lois Jones puts it, "when your bin is full and overflowing, you complain about the Station emphasis on production, but where would we be without it?"[13]

One argument made by advocates of production research is narrowly economic. They note that North Dakota is an agricultural state and that production research helps it maintain a favorable position relative to competitors here and abroad. "I don't think you can ever get too efficient," former cattleman Frank Kubik argues. "If you can produce the same amount with smaller acreage . . . what's the matter with that?" Or, as Bryce Streibel bluntly puts it, "if we can get 100 bushels an acre *great*—our competitors are trying to do it."[14]

Others see production research as necessary for maintaining food security and for meeting America's moral responsibilities to the world. Governor George Sinner, for example, worries that we are "living on borrowed time with world food supplies....There have been some very scary periods in world food reserves....This whole thing can change overnight. You can't just look at this year or five years or ten years." The devastating drought that hit the Great Plains and Midwest in 1988 and the consequent drawdown in world food reserves imparted a special relevance to Sinner's remarks.[15]

As long as the NDAES remains an institution with limited flexibility and as long as the essence of agricultural science remains unchanged and as long as North Dakota lives by producing and selling agricultural products, the Station will be heavily involved in production research. Some of this will simply be maintenance research designed to address immediate production problems. This sort of work is not very controversial, because no farmer, whether prices are low or high, and whether he or she is farming the government or operating in a free market, desires to produce less than can be produced or to produce crops and animals of lower quality. On the other hand, some production research will involve the search for breakthroughs that could result in dramatic increases in production in the future. This research will always be controversial, and at times it will seem inappropriate or even harmful to producers. It is preparation for a situation that may never occur, and much of it will never bear fruit, but it will be done for the foreseeable future.

The fact that there will always be production research does not mean that the nature of production research will remain the same. In the past few years the Station has shifted considerable resources into biotechnological research. Any major breakthroughs in production in the next generation will probably come from biotechnologists.

Biotechnologists study plants and animals at the cellular and subcellular levels. Using biotechnological techniques, it is feasible for researchers to grow plants from single cells, study individual cells for such traits as disease resistance, and fuse two cells from different plants to form a new plant. At the subcellular level it may become possible to alter the genetic structure of a plant or animal in order to give it characteristics it does not have. Biotechnologists have also succeeded in creating new bacterial life forms that may have a variety of agricultural applications.

Many observers of agricultural research are enthusiastic about the possibilities of biotechnology. Weed scientist John Nalewaja, for example, believes "biotech might help us incorporate resistance, or help a plant produce a chemical which will prevent a weed from growing." Keith Bjerke is even more optimistic, contending that, because of advances in biotechnology, "we'll be able to design plants . . . . We will have nitrogen fixation in small grains . . . . We will produce plants with their own resistance to diseases and insects, and they will probably be area specific."[16]

By its very nature, biotechnological research has both applied and basic aspects. On the applied level, biotechnology allows much more rapid development of new varieties because it enables the researcher to join single cells to one another and thus skip the sexual crossing that plant breeders have traditionally employed. Plant pathologist Gary Secor believes this makes biotechnology especially applicable to problems with which the Station has traditionally dealt. Biotechnology and plant breeding have "a perfect marriage," Secor contends, "because we can produce germplasm to be used in breeding programs."[17]

Subcellular biotechnological research is more basic, as it must be until we learn more about genetic structure and its manipulation. Of course, basic research often results in applications. L. R. Waldron embarked on wheat breeding because he wanted to learn about genetic inheritance, but the products of his work benefited North Dakota farmers in a very tangible way.

There are those who justifiably wonder whether the legislature will be willing to fund work that does not promise a clear and immediate payoff. Longtime GNDA official and Station observer Richard Crockett notes that "the state won't support basic research as well as they support meat and potatoes research. Biotech looks good in the newspaper, but you need to see it on your farm." Clark Jenkins adds that "it's easier to sell something you can see than something that's hypothetical." Even at best, funding in North Dakota is not liberal, leading some observers to question the potential of the NDAES in biotechnological research. "Biotechnology is a long shot," Station veteran Howard Olson believes. "I have to wonder if we won't have to be very lucky at SU to discover anything very important, given our funding disadvantages."[18]

Some Station veterans doubt that biotechnology will live up to its billing. Robert Johansen, for example, believes "it's overplayed" because "so many of the things they talk about haven't happened yet." Agricultural engineer George Pratt sagely notes that "a technology can be successful in the test tube and not in the fields." Jack Carter adds that "biotech isn't going to give us a lot, and it's not going to come very soon." While conceding that "biotechnology has its place," Tom Conlon believes it is largely "a buzz word cleverly contrived to keep the politicians confused."[19]

This attitude is probably informed by several considerations. Foremost is the wisdom gained from decades of work in agricultural research. Every few years a development or technology comes along that promises to revolutionize agriculture and agricultural science, but many of these have proved to be disappointing. It may also be the case that older scientists feel threatened by a development they do not fully understand. Richard Kiesling notes that "everything I learned in graduate school in the late forties and early fifties has been disproven." The sort of fundamental scientific shift that biotechnology represents can be very frightening to people trained when now-discarded assumptions and techniques were in vogue. And it may be that some researchers—traditional plant breeders, for example—fear they will become little more than technicians, carrying out the instructions of biotechnologists. Status is an important consideration in all human institutions and relationships, and change inevitably threatens to rearrange the status of individuals.[20]

Adding to the challenge of biotechnology is the fact that it does violence to traditional disciplinary boundaries. While agricultural research is traditionally highly cooperative and interdisciplinary, biotechnology demands that these traits become even more prominent. As former associate Dean of Agriculture Ross Wilkinson, head of the Agricultural Biotechnology Center until 1989, noted in

1987, "I not only want to mix plant scientists together, but I want to mix them with animal scientists, too. The problem I have is the resistance to the breakdown of the disciplines." The attempt to meld disciplines together will result inevitably in some frictions, resentments, and turf battles.[21]

Another potential problem for biotechnological research lies in the realm of public opinion. There are those who question the creation of new life forms or the genetic alteration of existing forms on moral grounds; who believe, in essence, that man should not join together what God has put asunder. Others worry about the environmental implications of the use of some products of biotechnology in the field. A genetically engineered bacteria that retards frost in a particular crop, for instance, might have unforeseen deleterious ecological consequences. These concerns have not been much heard yet in North Dakota, but the Station needs to be aware of their existence.

Those engaged in biotechnological research have demonstrated a healthy respect for the risks involved in this new undertaking. They are taking what agribusinessman Richard Golberg calls "a rifle-shot approach," which concentrates on a handful of crops, rather than a "shotgun approach." The Station is determined to concentrate on North Dakota crops and look for practical applications that can be introduced fairly rapidly.[22]

The biotechnologists, like others at the Station, are also very careful to retain a close relationship with clients. "People in the state are very accepting of what we are doing—they don't necessarily understand it, but they are optimistic . . . ," Gary Secor believes. "A large part of all of our time is spent dealing with grower concerns—they support our research, but they expect us to be responsive to them." As biotechnological research unfolds at the NDAES it is likely to remain narrowly focused, client sensitive, and applicable to the special problems and conditions of North Dakota agriculture.[23]

Despite the many and various risks of biotechnological research, this is the direction in which the Station must move. Bryce Streibel believes that "we can't afford not to be on the cutting edge of research," and he is right for at least three related reasons. First, most of the gains facilitated by mechanization, chemicals, and conventional plant breeding have already been achieved. Future breakthroughs, if they come, will probably be brought by people in the life sciences dealing with genetic structure. If the NDAES is to continue to provide optimum service to North Dakota farmers it must do this kind of research. As NDSU president Jim Ozbun notes, the NDAES cannot ignore biotechnological research "if we want to maintain a competitive society in North Dakota." The second reason why the new emphasis is necessary is because that is where the money is. Competitive grants from the federal government and agribusiness are targeted at biotechnological research. A final consideration involves the maintenance of a reputable scientific institution. Good young scientists today will not be attracted to or held in institutions that are not doing cutting-edge research. The NDAES could not have held L. R. Waldron if he had been expected to do regulatory work, and it will not be able to hold quality scientists in the future if it does not move with the times. Virtually all observers agree that the future of agricultural science in the United States lies in genetic research. If the NDAES is to remain a reputable scientific institution and provide maximum service to the state it must be part of that future.[24]

II

The new emphasis on biotechnology at the NDAES raises again the chronic problem of funding. Through most of Roald Lund's tenure, as was the case during much of Arlon Hazen's directorship, legislative appropriations have barely kept pace with inflation, though the Station has usually fared better than most state institutions. Former governor William Guy bemoans the fact that Station scientists are "the most important people in the state and the most poorly paid" and that they must "continually beg to keep feeding the goose . . . laying the golden eggs." At the same time, funding from the United States government has generally lagged behind inflation and has become relatively less significant. Moreover, appropriations from Washington come with ever more strings attached to them.[25]

In North Dakota a number of factors have resulted in lagging legislative appropriations. There is, first and foremost, the reality that North Dakota is a poor state, and a poor state that has suffered during the eighties from depressed conditions in both agriculture and energy. In times of budget stringency legislators strive first to meet short-term needs, sometimes slighting long-term programs such as agricultural research, especially when these provide relatively intangible payoffs. There is also the problem that the most devoted constituency of agricultural experimentation is shrinking. Fifty years ago there were about eighty thousand farms in North Dakota; today there are barely thirty thousand, and most North Dakotans live in cities. Not all farmers are committed to agricultural research, of course, especially when overproduction is a problem, but they tend to be stronger supporters than nonfarmers. The Station argues vigorously that all North Dakotans benefit from its work. But, it is easier for the farmer to see how a new potato variety benefits him or her than it is for the town physician, business person, teacher, or worker.

Constricted budgets have made it difficult for the NDAES to serve the public as fully as it would like. Travel and library funds are thin, and state-of-the-art equipment often lies beyond the Station pocketbook. Hiring and keeping quality people at a time when the supply of newly trained scientists is not abundant has been difficult. This is an especially crucial problem now, when key station veterans are retiring and the new biotechnological emphasis is being put in place. Compounding this difficulty is the fact that agricultural scientists are highly coveted by agribusiness. Some researchers have been offered as much as double their present salaries by chemical, seed, malting, and food processing concerns. Many have stayed because they like the area and the people, enjoy the University and Station atmosphere, and appreciate the ability to work with students. Still, one's duty to self and family sometimes dictates sacrificing these satisfactions.

In 1987 the state's mushrooming budget crisis forced dramatic cuts in Station funding. Bryce Streibel believed the reductions were "penny wise and pound foolish" because "research is one of the greatest legacies we can leave to future generations," but the legislature was simply trying to negotiate a difficult biennium. At the NDAES the cuts resulted in a number of early retirements and in severe reductions in several departments. Agricultural Engineering, Biochemistry, and Soil Science were hit especially hard.[26]

The budget stringency of the eighties led to some healthy examination of the research system and its component parts. Some wondered whether such areas as agricultural engineering, entomology, and weed science could be privatized. Others questioned whether all—or any—of the substations were necessary and whether horticultural research outside of potato breeding could really be justified.

Privatization was a popular buzz word in the United States during the conservative, antigovernment eighties. Because most Station scientists are conservative people who favor a limited social and economic role for government, they give a good deal of thought to the utility of surrendering some of their functions to the private sector. Most agricultural researchers see two main problems with privatization. First, they question whether private entities would be interested in limited market opportunities, such as those presented by special problems suffered by only a handful of farmers. As agricultural engineer George Pratt notes, "private industry is interested only in potential profitability—we fill the gaps, dealing with specialized crops and areas of doubtful profitability." This is related to the second problem, that of purpose. The purpose of the public system is to serve the clients by addressing their specific short- and long-term needs. The purpose of private businesses is to profit. Certainly, private enterprise may and often does serve the public in its quest for profit, but such service is incidental to its main goal. It is hard to imagine that the individual farmer with a weed or insect problem would receive from Monsanto or Velsicol or any other major agribusiness entity anything like the sort of service he or she receives—and expects—from the Station.[27]

Research centers—called branch stations before the 1989 legislature authorized a name change—are another area where cuts might be made. North Dakota has eight of them, at Carrington, Dickinson, Hettinger, Langdon, Mandan (the Land Reclamation Research Center), Minot (the North Central Research Center), Streeter (the Central Grasslands Research Center), and Williston. Supporting these entities has been expensive and more than one director has seen them as more bane than benefit. Nevertheless, the research centers provide a number of important and tangible benefits for the research system and the state.

They make research visible throughout North Dakota and help the NDAES retain broad political support. Each has a distinct, specialized role, and the branches provide a range of environmental and climatological regimes that facilitate variety testing and seed increase. Having a substantial number of centers allows the NDAES to take potentially beneficial risks, such as the devotion of the Carrington facility to research on sustainable agriculture. The research centers are most significant as providers of important services to local farmers and ranchers. Tom Conlon, superintendent of the Dickinson Research Center, notes that "farmers are often friendlier to the branch stations, because they need immediate help . . . . You live with the farmers and you're part of the farmers' lives." Longtime Carrington superintendent Howard Olson adds that the branch stations "work with the grassroots people and their real problems . . . . The farmers learn from us, and we learn from them." Many farmers apparently feel strongly attached to the research centers. Of the Dickinson facility Frank Kubik says, "they're down to earth—everybody feels comfortable with them. You can take any problem to them."[28]

Horticultural research is another area that seems unnecessary to some people, because much of it deals with aesthetics rather than dollars and cents. There probably is not much of a future for commercial squash or tomato growing in North Dakota, but how does one calculate the economic—or psychic—value of the backyard gardens of the state? The development and testing of appropriate trees is an important function in a state in which shelterbelts preserve the soil and shade trees enrich the soul. The benefits of such research are not always easily detected on balance sheets, but that does not mean it does not improve our lives.

The purpose of the foregoing discussion is not to suggest that every portion of the NDAES is indispensable to the survival of North Dakota agriculture. There is no doubt that people would still grow crops and raise animals in the state even if no research facility of any kind existed. Nor would anyone argue that the system is or should be set in stone. Departments have come and gone, and functions have been added and dropped over the last century, and there is no reason to believe that process will cease. The system must change to avoid the sort of ossification about which Myron Just worries. But there is no portion of the system today that is clearly superfluous or dysfunctional. Each part of the system plays a distinct role and provides benefits to North Dakotans. The loss of any portion would diminish the degree to which the system as a whole could serve the people of the state. The NDAES does well and has always done well for the state with the resources it has available.

Begging from a beleaguered legislature is not the only way to raise funds. The NDAES can increase its resources by more aggressively pursuing profits, and a promising route to the generation of more capital—and one that is being followed by a number of experiment stations in the United States—is the patenting of new techniques or genetic material. This would profit the Station because once a new variety, for example, was patented, users would have to pay royalties, just as users of industrial devices pay royalties to patent holders or users of copyrighted material pay royalties to the owners of the copyrights.

The potential drawbacks of patenting touch on the bases of the Station and its purposes. As a public scientific institution the NDAES believes implicitly in the advancement of knowledge through the open interchange of information. The pursuit of patents and royalties would inject an element of competitiveness and secrecy into the scientific enterprise. Patenting "may lead to squirreling of information and material," Richard Kiesling points out. "It's very harmful to the free exchange of information in academics." Patenting would probably also force an alteration in the long-established Station policy of getting new varieties to producers as quickly and cheaply as possible. Directly or indirectly, patenting would likely drive up the prices clients pay for Station services.[29]

Another potential route to enhanced funding lies in developing closer relationships with agribusiness. The NDAES has never shunned corporate money. In the early years such businesses as the railroads and the paint and varnish manufacturers provided support. Later, chemical companies, millers and bakers, and industry groups such as the American Malting Barley Association made important contributions. The NDAES has always been grateful for supplementary funds, but it has also been cautious in accepting them. Insofar as private funds could be seen as contributing to the Station's purpose of serving the state and its farmers they were accepted. Insofar as private monies potentially compromised the Station's public purpose they were declined.

The balance between private support and public responsibility remains as delicate as it ever was. Richard Crockett believes that "private enterprise contributions are fine—they don't buy you just because they give you money." There is much to what Crockett says, but through its unquestioned right to give money when and where it wishes agribusiness can shape research emphases and directions. As Jim Ozbun puts it, "private support restricts our capacity to do what we want to do . . . . State money allows the scientist to be creative."[30]

At present agribusiness is especially interested in biotechnological research. That interest, which reflects the corporate perception of where the major breakthroughs will be in the future, is good for the Station, because biotechnological research is expensive and much of it is of the basic type that legislators are reluctant to fund. Such funding carries with it the possibility of future financial well-being, both because agribusiness firms will throw good money after good and because it might result in patentable developments from which the Station might profit.

Some observers of the publicly supported agricultural science system see potential problems with agribusiness patronage of biotechnological research. Many at the NDAES share these concerns. Obviously, agribusiness would like the stations to do basic research that can then be applied by the private sector in the form of new products, such as plant varieties. The growing involvement of agribusiness firms in variety development raises some issues that must be addressed. One is the issue of cost. Gary Secor believes that "if private companies produce new varieties, they'll be twice as expensive" for the farmers as varieties produced by the Station. The possibility also exists that the NDAES might be foreclosed altogether from producing new varieties, either through the patenting process or through contractual obligations imposed on the Station by agribusiness grantors. That would be troubling because, as Bob Johansen notes, "our reputation is based on the number of varieties we've introduced." Not only would the Station lose visibility if it no longer introduced varieties, but it would also potentially lose political support. Orville Banasik contends that "it would be difficult to sell the Station to the Legislature if it didn't develop new varieties," and most Station veterans agree. Declining legislative support would likely lead to increasing dependence on agribusiness and a diminished capacity to address directly the needs of North Dakota farmers. Some thus see in corporate support the potential for the capture of the NDAES away from those it was created to serve. When Roald Lund says that "once you cross the bridge away from state support you never go back," he is pinpointing a serious potential problem.[31]

Most Station scientists believe that North Dakota has been well served by the system of predominant state funding. But it is the citizens who will ultimately determine how they will be served and at what level. It is they who will decide whether there should be more or less public support and more or less private support. The Station will live, as it always has, with the verdict of the state.

III

Many of the challenges the NDAES faces as it contemplates the next century are not really new. The question of how best to serve the public, the issue of which scientific direction to take, the constant struggle for political support, the chronic worries about funding, the nature of the relationship between public agencies and private entities—not one of these is a fresh concern at the Station. Were the shade of Horace Stockbridge, or John Henry Worst, or Edwin Ladd to appear in Roald Lund's office, it would be mystified by the details of modern science, but it would grasp the institutional realities of the NDAES immediately.

Another thing that has not changed much is the value system agricultural researchers at the NDAES manifest. In significant ways they are nineteenth-century people, sharing the values of those "Whig-Republicans" who created the system more than a century ago. They are believers in human possibility. They remain certain that individuals can advance themselves and society through education. They believe that science is the handmaiden of progress and that social as well as technical problems can be solved by it. They seek to improve life for children and grandchildren as much as for self. In short, they have kept the faith so many of us have lost.

The personal behavior of so many Station scientists also reflects the values of a bygone age. They are people of integrity. They believe in hard work, self-restraint, and self-reliance. Most of them continue to hold, as Lincoln held and Jefferson before him, that government should do for people only what they cannot do for themselves. And they are conscious that they work for the people and are as careful with the public dollar as they are with their own.

Though they consider themselves practical people, Station researchers are idealists. In many cases they were drawn to agricultural science because they believed, in Chris Jensen's words, that it was "a service to mankind." And they stayed in it and found satisfaction in it because it allowed them to satisfy their ideals. When Bob Johansen says that he is gratified to have produced varieties that have "improved people's lives," it is clear that he is in the research enterprise because it allows him to serve others.[32]

In clinging to these articles of faith agricultural scientists find themselves increasingly out of step with American society. The notion that the individual is good and can improve seems archaic and wrongheaded in a century in which people regularly perpetrate horrors of the most beastly sort. Science for some is more an object of fear than an article of faith; for others it is mainly a means to make a dollar or obliterate an enemy. In a society in which people think mainly of the short term and the bottom line, in which self-indulgence and ethical flexibility reign, in which most wish to take no risks and bear no blame, and in which idealism is viewed with cynical distrust or bemusement, the agricultural scientists seem out of place. For most of its history the system has enjoyed the assurance that it was helping society reach its goals and realize its ideals. It no longer has that assurance because our society no longer has a clear set of goals and ideals. That problem, as much as any other, clouds the future of the agricultural research system.

But the problem of values, significant as it is, represents only one of the clouds on the horizon. Another is the incessant and insistent criticism of the quality of experiment station science from impatient scientific elitists. Still another is the basic shift in the United States from a production to a service economy. Agricultural research is oriented toward producers—people who create tangible products to fulfill basic human needs. Whether such an enterprise will remain relevant in an economy of burger flipping and financial consulting is an open question.

Even within the shrinking production economy, agriculture holds an ever-smaller place. In part, ironically, because of the successes of the research enterprise, fewer farmers are needed to produce enough to fill America's domestic needs and its export requirements. As farmers become less important economically and politically the research system becomes more superfluous and vulnerable. In a number of states already experiment stations have become trivial and anachronistic remnants of a bygone age.

The foreseeable future does not hold that fate for the NDAES. North Dakota remains a producer state and an agricultural state. Still, even North

Dakota has fewer farmers every year. What was once the North Dakota Agricultural College has been North Dakota State University for more than a quarter of a century. Within that institution the College of Agriculture serves a minority of students that seems to dwindle a bit nearly every year. In 1988 a new university president was hired to replace L. D. Loftsgard, who died in October of 1987. The intense desire of many in the Station and the College of Agriculture that the new president have a background in the agricultural sciences was perhaps an indication of anxiety about the future.

The fact is that, while it has been the agent of great change and has undergone great internal alterations, the agricultural research system has been characterized by essential continuity. In so many essential areas—its values and goals, its client base, the direction it has charted for itself and for agriculture—the system has not changed very much.

But if continuity is in one sense a weakness of the system it is in another sense its strength. It is reassuring to see programs that endure for generations, continuing commitments to goals, and unaltered devotion to service.

In its continuity, as well as in its emphasis on cooperation, self-sacrifice, and on handling problems internally, the NDAES resembles a family. Associations within the Station are at once professional and social. Scientists work and play together, celebrate and grieve together, and mark the important occasions of their personal and professional lives together. They share so much in terms of training, outlook, and even background that it would be curious if they acted otherwise.

This atmosphere is not for everybody. Some find it oppressive, and it contributes to the tendency to insularity many perceive in agricultural research institutions. But most of the people in the NDAES find the familial environment secure and comforting. They enjoy a sense of belonging rare even in the academic community.

There is a generational structure to the NDAES that is also akin to that in a large and extended family. Young people come in, often as undergraduate students, spend their careers in the Station, and continue to interact with younger colleagues as respected professors emeriti after retirement. Ted Stoa knew every president from Worst through Loftsgard. O. A. Stevens inspected seeds for Bolley and lived to see Roald Lund in the director's office. Glenn Smith worked for L. R. Waldron at Dickinson and now observes the unfolding of biotechnological research at the Station. Range scientist Warren Whitman had Herbert Hanson as a mentor and now plays that role for a whole new generation of young researchers. Generational shocks are rare as individuals move smoothly from student to protege to mentor to respected elder statesman. Just now the post-World War II generation is retiring, passing the reins to Station baby boomers without bitterness or rancor. Retirements are major events at the NDAES. They provide an opportunity to show appreciation and respect. They are not occasions to see retirees off or to consign them to oblivion, because the Station continues to attend to them and celebrate them for as long as they live.

The flow and continuity of lives and careers at the Station imparts a durable—almost eternal—quality to it that is remarkable for an institution only a century old. Chris Jensen once said of James R. Dice, who came in 1920, that "we expected him to be here a long time, but he died in 1945." That comment seemed humorous to me, because in our transient society with its ephemeral attachments twenty-five years seems like a long time. But in an institution like the NDAES, to which many devote most of their lives, a quarter century is not so long.[33]

The familial sense and the devotion it entails extends beyond the Station to the state. Scientists work with farmers and their sons and daughters for decades. Substation superintendents become integral parts of local communities. Sons and daughters of farmers' neighbors take Station posts from which they can help farmers' sons and daughters. The devotion of researchers to the state is unaffected, and it is repaid with affection. Scientists and clients are bound together by mutual respect and by the ties created by years of working together in a common enterprise. The NDAES is a scientific institution driven by a human dynamic.

The fact that the NDAES is a living, human institution rather than a mere line appropriation limits the degree to which it can change. It is a product of its history, which its people carry within them, and its future will unfold from its past. There is more good than bad in that fact. The best thing about it is that the state can count on a commitment to service at the NDAES. In the first annual report Horace Stockbridge told his readers that "we wish to impress upon the attention of the farmers of the State that this institution belongs exclusively to them, that it receives its income for the sole purpose of being expended in their aid, [and] that in its work their desires and interests will be constantly furthered." In 1987 Roald Lund said that "all of the technical expertise in the world can't compensate for a commitment to service. The people in the state tell us how they want us to serve them, and we do it." In our rapidly changing world, there is much to be said for such constant devotion to a high ideal.[34]

History can predict neither what the problems of North Dakota agriculture will be in a hundred years nor how North Dakota farmers will want the Station to serve them. But history does tell us enough about the character of that institution to give us a good idea how it will respond. As long as there is a need and as long as it has the means, the NDAES will serve the public. That is its past. That is its future. That is its nature.

THE NORTH DAKOTA AGRICULTURAL EXPERIMENT STATION IN THE CENTENNIAL YEAR

Agricultural Administration
First row, left to right: Margaret Olson, Tammy Dahl, Sharon Morgan, Diane Tronson, Mary Wiger
Second row, left to right: Margot Peterka, Rovert Harrold
Third row, left to right: H. Roald Lund, Don E. Anderson, Jim Swank, Jeff Halvorson, Dave Kneip, Tom Lewis, Harvey Vreugdenhil, Greg Sauvageau, Don Hanson

Agricultural Economics
First row, left to right: Carol VavRosky, Cindy Schuler, Shelly Swandal
Second row, left to right: Marna Unterseher, Carol Jensen, Judy Moe, Brenda Eckstrom
Third row, left to right: David Cobia, Won Koo, Jerome Johnson, Cole Gustafson, Delmer Helgeson, F. Larry Leistritz, Thor Hertsgaard, Richard Rathge, David Watt, Roger Johnson, Randall Sell, Dean Bangsund, Timothy Mortensen, Steven Gunn, Lawrence Mack

Agricultural Engineering
Left to right: Leslie Backer, Charles Moilanen, Dennis Wiesenborn, James Lindley, Mary Hanzel, Tom Bon, Margaret (Peg) Odegaard, Elton Solseng

Animal and Range Sciences
First row, left to right: Clayton Haugse, Dana Millaway, Trent Gilbery, Marsha Kapphahn, Deanne Morse, LaDon Johnson, Chung Park, Ronald Zimprich, Anna Grazul-Bilska, Julie Berg, Arlinda Lewis, Wanda Keller, Cindy Belohlavek, Gloria Ottem
Second row, left to right: Mark Scheetz, James Kirsch, Gerald Erickson, Kim Kraft, Gregory LaCroix, William Slanger, Kevin Sedivec, Robert Weigl, Russ Danielson, Stephen Boyles, Bert Moore, David Zaeske, Dale Redmer, Thomas Schmaltz, Lawrence Reynolds, Royal Carlson, Timothy Flakoll, Robert Harrold, Beverly Liebelt, Joe Crenshaw, Sharon Vigesaa, William Barker, Mario Biondini, Duane Erickson, James Tilton, Robert Johnson

Cereal Science and Food Technology
First row, left to right: Navam Hettiarachchy, Paul Schwarz, Dehdra Puhr, Angela Ostenson, Elaine Hanson, Cindy Bruns, Yvonne Holm, Andrea Ottenbacher
Second row, left to right: Clarence McDonald, Linda Huckle, Sharon Swanson, Linda Grant, Khalil Khan, Rachel Nelson, Vernon Youngs, John Jansky, Lori Baumann
Third row, left to right: Bert D'Appolonia, Evan Cummings, George Matthiensen, Dale Hansen, Truman Olson

Crop and Weed Sciences
First row, left to right: Chao-Chien Jan, Albert Schneiter, LeRoy Spilde, Harold Cross, Murray Duysen, Marilyn Schroeder, Elaine Dobrinz, Iris Bachand, Edward Deckard, Leonard Joppa, Norman Williams, Calvin Messersmith, A. Earl Foster
Second row, left to right: Burton Johnson, James Hanzel, Dale Rehder, Jerry Miller, Ron Roach, Frank Manthey, Mark Ciernia, Roy Cantrell, Amy Jo Johnson, Marilyn Geiszler, Mardell Krebsbach, Donald Galitz, William Ahrens
Third row, left to right: Mark Halvorson, Garry Mowery, Gerald Johnson, Ken Grafton, Kay Carlson, Ted Helms, Rich Horsley, Phil McClean, Carl Madsen, Bob Baumann, Jim Faller, Josef Maianu, Eileen Buringrud, Jody VanderWal, Roger Genoch, Jerome Franckowiak

Entomology
First row, left to right: Donald Hyder, Tariq Mahmood, Gerald Fauske, Sandra Osborne, Teresa Gross, Barbara Bates, Ronda Conlon, Joanne Thoemke, Teryl Grosz, Colwell Cook, Ted Schulz, Dean Christie, Jeff Teerink
Second row, left to right: Gary Brewer, Larry Charlet, Mark Johnson, Joy White, Don Mundel, Robert Carlson, Chris Oseto, Ed Balsbaugh
Third row, left to right: Mark Anderson, Graig Roseland, Walt Valovage, Don Carey, Albin Anderson, Dean McBride, Wilfred Burr, David Sether, Mike Weiss, Hendrik Meyer

Horticulture and Forestry Department
First row, left to right: Art Boe, Robert Johansen, Mark Ehlenfeldt, Dale Herman, Louise Heinz, Luella Carlson
Second row, left to right: Delila Thompson, Robin Conlon, Dean Peterson, Kristie Wolff, Pam Loose, Shelley Jansky
Third row, left to right: Jerry Tuskan, Bryce Farnsworth, Myron Thoreson, John Lundquist, Wayne Sargent, Larry Chaput

Plant Pathology
First row, left to right: Glen Statler, Martin Draper, Raymond Taylor, James Jordahl, Paul Henningson, Chris Onstad, Devia Muccatira, Basilio Salas, Freshtah Saghaei, Yu Zhi Zheng
Second row, left to right: Annette Pfeifer, Renee McClean, Diana Iverson, Lillian Hjelmstad, Roberta Haspel, Marcia McMullen, Vicki Gustafson, Debra Baer, Thomas Gulya, Phyllis Hellem, Tracy Christianson, John Eschbach, Katherine Warner, Hsiao-ling Wu
Third row, left to right: Ronald Koech, Robin Lamppa, Thomas Fetch, Roy Haug, Phillip Nolte, James Miller, James Walla, Berlin Nelson, James Venette, Ilhan Kural
Fourth row, left to right: Neil Gudmestad, H. Arthur Lamey, Gary Secor, Mark LeRud, Robert MacArthur, Michael Edwards
Fifth row: Kerman Alm

Soil Science
First row, left to right: Lynn Brun, Marjorie Dobervich, Sharon Monson, Trudy Lane, Arlyce Johnson, Edward Deibert, Raymond Knighton
Second row, left to right: Donald Patterson, Dean Lizotte, William Dahnke, John Enz, Jay Goos, Brian Johnson, Lyle Prunty
Third row, left to right: David Hopkins, Chris Brenk, Rodney Utter, Larry Swenson, Joseph Giles, Michael Sweeney, Jimmie Richardson, Larry Cihacek

Microbiology/Veterinary Science
First row, left to right: M.H. Smith, David Berryhill, Traci Spafford, Evelyn Lundeen, Virginia Leverson, Dawn Doetkott, Linda Parmer, Deanne Iverson, Beth Tacke, David Khansari
Second row, left to right: Heidi Gorrilla, Mary Moen, Darlene Krogh, Sylvia Kistner, Vicki Rapp-Gabrielson, Joann Colville, Beverly Baldwin, Jim Struble, Kevin Soiseth, Teresa Sonstahagen, Eloyes Hill, Howard Casper
Third row, left to right: Marvin Ban Cante, Melbert Fisher, Charley Koesterman, Thomas Glass, Thomas Colville, David Gabrielson, George Schamber, Arnold Alstad, Eugene Berry, Alan Misek
Fourth row, left to right: Terry Lewis, Michael Saxerud, Ivan Berg, Thomas Gustad, Berdell Funke

Agronomy Seed Farm
Left to right: Curt Vesey, Lloyde Peck, Tom Teigen

Carrington Research Center
Standing left to right: Brad Miller, Dave Copenhaver, Steve Zwinger, Steve Guldan, Betty Montgomery, Caroline Wolf, Yvonne Larson, John Burbank, Blain Schatz, Vern Anderson, John Gardner
Kneeling left to right: Dale Burr, Marvin Hollatz, Patrick Carr

Central Grasslands Research Center
Left to right: Brian Kreft, Jim Nienow, Bob Patton, Anny Nyren, Paul Nyren, Al Scheet

Dickinson Experiment Station
Left to right: James Nelson, David Claypool, Douglas Landblom, Thomas Conlon, Jane Olheiser, Ann Bertie

Hettinger Research Center
Left to right: Dan Nudell, Cave Pearson, Nels Olson, Don Stecher, Terri Lindquist, Shari Wick, Kris Ringwall, Eric Eriksmoen, Phil Berg, Tim Faller, Kate (dog)

Land Reclamation Research Center
First row, left to right: Kevin Vining, F. Scott Carter, Padam Sharma, Michelle Miller, Shelly Muller, Linda Thomas, Jodi Koenig
Second Row, left to right: Steve Schroeder, Steve Nelson, Kelvin Kosse, Gary Halvorson, Zann Nash, Karen Stastny, LeRoy Zimmerman, Eugene Doll

Langdon Research Center
Left to right: John Lukach, Sharon Ring, Dick Wilhelmi, Bryan Hanson, Kevin Beauchamp, Jim Chaput, Ron Stover

North Central Research Center
First row, left to right: Ben Hoag, Jerry Hennessy
Second row, left to right: Llauren Nelson, Lori Sundsbak, Curtis Thompson

Williston Research Center
Left to right: Ernest French (Ernie), Gordon Bradbury, Marilyn Krogen, Neil Riveland, Richard Helgeson, Ted Krogen

NOTES:
CHAPTER 1
BEFORE THE BEGINNING

1. William C. Hunter, *Beacon Across the Prairie: North Dakota's Land-Grant College* (Fargo: North Dakota Institute for Regional Studies, 1961), 19.
2. For federal sponsorship of scientific agriculture see A. Hunter Dupree, *Science in the Federal Government: A History of Policies and Activities to 1940* (Cambridge: Harvard University Press, 1957); and Norwood Allen Kerr, *The Legacy: A Centennial History of the State Agricultural Experiment Stations, 1887-1987* (Columbia: Missouri Agricultural Experiment Station, 1987). For the early development of American experiment stations see Alan I Marcus, *Agricultural Science and the Quest for Legitimacy: Farmers, Agricultural Colleges, and Experiment Stations, 1870-1890* (Ames: Iowa State University Press, 1985). Much of the material and many of the conclusions that follow regarding the foundation of the experiment station system are derived from my chapter, "Publicly-Sponsored Agricultural Research in the United States from An Historical Perspective," in Kenneth A. Dahlberg, ed., *New Directions for Agriculture and Agricultural Research: Neglected Dimensions and Emerging Alternatives* (Totowa, New Jersey: Rowman and Allanheld, 1986), 107-31.
3. For the professional ideal in nineteenth-century America see Barton Bledstein, *The Culture of Professionalism: The Middle Class and the Development of Higher Education in America* (New York: Norton, 1976). For its specific application to agriculture see Charles Rosenberg, "Rationalization and Reality in the Shaping of American Agricultural Research," *Social Studies of Science* 7 (1977): 401-22.
4. W. O. Atwater, "The What and Why of Agricultural Experiment Stations," *Science* 14 (August 9, 1889): 97.
5. A few of the major works on North Dakota's geology and climate include John H. Hainer, "The Geology of North Dakota," *North Dakota Geological*

Survey Bulletin 31 (1956); Daniel E. Willard, *The Story of the Prairies, or, The Landscape Geology of North Dakota* (Kalamazoo, Mich.: Ihling Bros. Everard Co., 1921); H. L. Walster, "The Soils of North Dakota," *Quarterly Journal of the University of North Dakota* 14 (January 1924); and Frank J. Bavendick, *Climate and Weather in North Dakota* (Bismarck: U.S. Weather Bureau, 1946).

6. John Wesley Powell, *Report on the Lands of the Arid Region of the United States*, 2d sess., Exec. Doc. 73, House of Representatives (Washington, D.C.: Government Printing Office, 1878), 50-51.
7. For the bonanza farms see Hiram Drache, *The Day of the Bonanza* (Fargo: North Dakota Institute for Regional Studies, 1964); and Stanley Norman Murray, *The Valley Comes of Age: A History of Agriculture in the Valley of the Red River of the North, 1812-1920* (Fargo: North Dakota Institute for Regional Studies, 1967). Robert P. Wilkins and Wynona Huchette Wilkins note in *North Dakota: A Bicentennial History* (New York: Norton, 1977) that the garden-desert controversy regarding the nature of the Great Plains went on through most of the nineteenth century.
8. My understanding of the North Dakota boom is derived mainly from Elwyn B. Robinson's *History of North Dakota* (Lincoln: University of Nebraska Press, 1966), 133-55.
9. Richard A. Bartlett, *The New Country: A Social History of the American Frontier, 1776-1890* (New York: Oxford University Press, 1974), 431; Howard W. Ottoson, Eleanor Birch, Philip A. Henderson, and A.H. Anderson, *Land and People in the Northern Plains Transition Area* (Lincoln: University of Nebraska Press, 1966), 37 and 47.
10. Wilkins and Wilkins, *North Dakota*, 84-85.
11. For the state of agricultural science see Margaret W. Rossiter, "The Organization of the Agricultural Sciences," in Alexandra Oleson and John Voss, eds., *The Organization of Knowledge in Modern America, 1860-1920* (Baltimore: Johns Hopkins University Press, 1979), 211-48.
12. W. O. Atwater, "Report of the Director of the Office of Experiment Stations," *First Report of the Secretary of Agriculture, 1889* (Washington, D.C.: Government Printing Office, 1889): 543.
13. Hunter, *Beacon Across the Prairie*, 10.
14. Quoted by Hunter, *Beacon Across the Prairie*, 13.

CHAPTER 2

EARLY DAYS

1. Remarks of Horace E. Stockbridge, *First Annual Report of the North Dakota*

Agricultural Experiment Station to the Governor, February 1, 1891 (Fargo: Daily Republican, 1891): 12.

2. For some brief biographical details regarding Stockbridge see William C. Hunter, *Beacon Across the Prairie: North Dakota's Land-Grant College* (Fargo: North Dakota Institute for Regional Studies, 1961), 21.
3. My understanding of the early figures at the NDAES is drawn largely from Bill G. Reid, *Five for the Land and Its People* (Fargo: North Dakota Institute for Regional Studies, 1989).
4. Ladd's pure-food work is explored in Ralph J. Kane's "Edwin Fremont Ladd: North Dakota's Pure-Food Crusader," (Master's thesis, University of North Dakota, 1960).
5. For Bolley's contributions see H. L. Walster, "North Dakota Honors Two Agricultural Scientists," North Dakota Agricultural Experiment Station, *Bimonthly Bulletin* 6 (July-August 1944): 4.
6. For example, the OES was unwilling to intervene when Stockbridge appealed to it during his struggle to wrest financial responsibility away from the Board of Directors. See, for example, his letter to Edwin Willets, September 12, 1891, Record Group 164, Office of Experiment Stations, State Correspondence, General (1888-1937) North Dakota, Oct. 1888-Dec. 1901, National Archives.
7. Remarks of H. L. Bolley, *Second Biennial Report of the Board of Trustees of the North Dakota Agricultural College to the Governor of North Dakota, 1893-94, Public Document No. 10* (Jamestown: Alert, 1894): 29.
8. Remarks of Horace E. Stockbridge, *Second Annual Report of the North Dakota Agricultural Experiment Station to the Governor, January 30, 1892* (Fargo: Walker Brothers, 1892): 5-10; and idem, *Third Annual Report of the North Dakota Agricultural Experiment Station to the Governor, February 1, 1893* (Grand Forks: Herald, 1893): 5.
9. W. O. Atwater, "Report of the Director of the Office of Experiment Stations," *Report of the Commissioner of Agriculture, 1888* (Washington, D.C.: Government Printing Office, 1889): 540; H. L. Bolley, "An Experiment; Its Conception and Methods of Procedure," *Proceedings of the Eleventh Annual Convention of the Association of American Agricultural Colleges and Experiment Stations, Held At Minneapolis, Minnesota, July 13-15, 1897, USDA Office of Experiment Stations Bulletin No. 49* (Washington, D.C.: Government Printing Office, 1898): 47; and idem, "The Status of Experimental Agriculture," *Science* 19 (May 13, 1892): 271.
10. Charles E. Rosenberg, "Rationalization and Reality in the Shaping of American Agricultural Research, 1875-1914," *Social Studies of Science* 7 (1977): 401. Also see my article "The Agricultural Experiment Station and Professionalization: Scientists' Goals for Agriculture, 1887-1910," *Agricultural History* 60 (Spring 1986): 246-55.
11. H. P. Armsby, "Agricultural Experiment Stations," *Popular Science Monthly*

23 (1884): 600.

12. A. W. Harris, "Report of the Director of the Office of Experiment Stations," *Report of the Secretary of Agriculture, 1891* (Washington, D.C.: Government Printing Office, 1892): 511.
13. W. A. Henry, "President's Address," *Proceedings of the Seventh Annual Convention of the Association of American Agricultural Colleges and Experiment Stations, Held At Chicago, Illinois, October 17-19, 1893, USDA Office of Experiment Stations Bulletin No. 20* (Washington, D.C.: Government Printing Office, 1894): 39. See also A. C. True and V. A. Clark, *The Agricultural Experiment Stations in the United States, USDA Office of Experiment Stations Bulletin No. 80* (Washington, D.C.: Government Printing Office, 1900); and Margaret W. Rossiter, "The Organization of the Agricultural Sciences," in Alexandra Oleson and John Voss, eds., *The Organization of Knowledge in Modern America, 1860-1920* (Baltimore: Johns Hopkins University Press, 1979), 211-48.
14. Remarks of Horace E. Stockbridge, *First Annual Report . . . Agricultural Experiment Station . . . 1891* (Fargo: Daily Republican, 1891): 11-12.
15. H. L. Bolley, Untitled autobiographical sketch, Box 2, "Bolley Biography" file, H. L. Bolley Papers, North Dakota Institute for Regional Studies; and remarks of Willet Hayes, *Third Annual Report . . . Agricultural Experiment Station . . . 1893*: 27.
16. *Yearbook of the United States Department of Agriculture, 1895* (Washington, D.C.: Government Printing Office, 1896): 526-35.
17. C. B. Waldron, "Some Profitable Vegetables for North Dakota," Government Agricultural Experiment Station for North Dakota, *Bulletin No. 18* (Fargo: March 1895): 121.
18. For dairy farming in North Dakota and the involvement of the NDAES therein see my article "The North Dakota Agricultural Experiment Station and the Struggle to Create a Dairy State," *Agricultural History* 63 (Spring 1989): 174-86.
19. Reid, *Five for the Land and Its People*, 23; and C. B. Waldron, "Tree Culture," Government Agriculture Experiment Station for North Dakota, *Bulletin No. 25* (Fargo: September 1896): 77.
20. Remarks of H. L. Bolley, *Fourth Annual Report of the North Dakota Agricultural Experiment Station to the Governor, February 1, 1894, Public Document No. 4* (Grand Forks: Herald, 1894): 9.
21. Remarks of Horace E. Stockbridge, *Second Annual Report . . . Agricultural Experiment Station . . . 1892*: 10.
22. Remarks of Horace E. Stockbridge, *First Annual Report . . . Agricultural Experiment Station . . . 1891*: 12; and idem, "Notice," in "Wheat Growing and Dairying for North Dakota," Government Agricultural Experiment Station for North Dakota, Bulletin No. 8 (Fargo: December 1892).
23. See the letter from Edwin F. Ladd to E. M. Upson, February 11, 1891, box 1,

folder 2, Edwin F. Ladd Family Papers, North Dakota Institute for Regional Studies. For a discussion of the methods of "research entrepreneurs" see Charles E. Rosenberg, "Science, Technology, and Economic Growth: The Case of the Agricultural Experiment Station Scientist, 1875-1914," *Agricultural History* 45 (January 1971): 1-20.

24. The following account of the political controversy at the Station synopsizes my article "The North Dakota Agricultural College Controversy of 1893: Scientific Professionalism and Political Patronage," *North Dakota History* 53 (Winter 1986): 12-23. The Independent movement is addressed in Elwyn B. Robinson's *History of North Dakota* (Lincoln: University of Nebraska Press, 1966); and in Glenn Lowell Brudvig's "The Farmers' Alliance and Populist Movement in North Dakota (1884-1896)" (Master's thesis, University of North Dakota, 1956).
25. Letter from E. F. Ladd to A. C. True, October 16, 1895, box 1, folder 2 Ladd Papers, NDIRS.
26. Letter from James B. Power to Andrew H. Burke, October 4, 1893, box 4, folder 1, letterpress p. 482, James B. Power Papers, North Dakota Institute for Regional Studies. The bonanza farm experiment, and Power's part in it, is recounted in Hiram Drache, *The Day of the Bonanza* (Fargo: North Dakota Institute for Regional Studies, 1964); and in Stanley Norman Murray, *The Valley Comes of Age: A History of Agriculture in the Valley of the Red River of the North, 1812-1920* (Fargo: North Dakota Institute for Regional Studies, 1967).
27. Remarks of Norman J. Colman, *Report of the Commissioner of Agriculture, 1888* (Washington, D.C.: Government Printing Office, 1889): 9; remarks of J. Sterling Morton, *Report of the Secretary of Agriculture, 1893* (Washington, D.C.: Government Printing Office, 1894): 9; and remarks of J. Sterling Morton, *Report of the Secretary of Agriculture, 1896* (Washington,D.C.: Government Printing Office, 1897): XLI. The USDA attained cabinet status in 1889, at which time its head ceased to be a commissioner and become a secretary.
28. A favorable assessment of Power's presidency is Stanley N. Murray's "James B. Power: The Second President of North Dakota Agricultural College," *North Dakota Quarterly* 42 (Autumn 1971): 5-25.
29. See the letter from James B. Power to Willet Hays, April 30, 1895, box 4, folder 2, letterpress page 152, Power Papers, NDIRS.

CHAPTER 3

BETTER DAYS

1. A. C. True, "Report of the Director of the Office of Experiment Stations,"

Report of the Secretary of Agriculture; Being Part of the Message and Document Communicated to the Two Houses of Congress At the Beginning of the Second Session of the Fifty-Fourth Congress (Washington, D.C.: Government Printing Office, 1896): 125.

2. Report from H. L. Bolley to J. H. Worst, December 31, 1913, box 9, "Reports, Various" file, p. 1, H.L. Bolley Papers, North Dakota Institute for Regional Studies.
3. For the second North Dakota boom see Elwyn B. Robinson, *History of North Dakota* (Lincoln: University of Nebraska Press, 1966), esp. 246-47. For North Dakota farm expansion see the *Fifth Biennial Report of the Commissioner of Agriculture and Labor to the Governor of North Dakota for the Two Years Ending June 30, 1898* (Bismarck: Tribune, 1898): 66, and the *Fourteenth Biennial Report of the Commissioner of Agriculture and Labor to the Governor of North Dakota for the Term Ending June 30, 1916* (Fargo: Walker Bros. and Hardy, 1916): 16. For national concern about agriculture in this period see my book *The Resisted Revolution: Urban America and the Industrialization of Agriculture, 1900-1930* (Ames: Iowa State University Press, 1979), esp. 36-42.
4. J. H. Worst, "Experiment Station Notes on Various Subjects," Government Agriculture Experiment Station for North Dakota, *Bulletin No. 31* (Fargo: March 1898): 247. For the Country Life Movement see my *Resisted Revolution*, esp. 42-74, and William L. Bowers, *The Country Life Movement in America, 1900-1920* (Port Washington, N.Y.: Kennikat, 1974).
5. J. H. Worst, *Eighth Annual Report of the North Dakota Agricultural Experiment Station, Agricultural College, N.D., to the Governor of North Dakota, February 1, 1898* (Bismarck: Tribune, 1898): 82; idem, *Seventh Annual Report of the North Dakota Agricultural Experiment Station, Fargo, N.D., to the Governor, February 1st, 1897* (Fargo: Argus, 1897): 46; idem, *Proceedings of the Twenty-Third Annual Convention of the Association of American Agricultural Colleges and Experiment Stations Held at Portland, Oregon, August 18-20, 1909, USDA Office of Experiment Stations Bulletin 228* (Washington, D.C.: Government Printing Office, 1910): 97, and idem, *Fourth Biennial Report of the Board of Trustees of the North Dakota Agricultural College to the Governor of North Dakota, 1897-1898* (Bismarck: Tribune, 1898): 22.
6. Remarks of Edwin Ladd and J. H. Worst, *Third Biennial Report of the Board of Trustees of the North Dakota Agricultural College to the Governor of North Dakota, 1895-96* (Bismarck: Tribune, 1896): 22 and 16. Some readers might find amusement in the fact that the building from which sewage was expelled by the use of hot air now houses the University administration.
7. Letters from A. C. True to J. H. Worst, October 2, 1897; and J. H. Worst to A. C. True, October 5, 1897, Record Group 164, Office of Experiment Stations, State Correspondence, General (1888-1937) North Dakota, Oct. 1888-Dec. 1901, National Archives.
8. Letters from J. H. Worst to A. C. True, April 1, 1899, E. F. Ladd to A. C.

True, March 31, 1899, and H. L. Bolley to A. C. True, April 3, 1899, OES, North Dakota, Oct. 1888-Dec. 1901, National Archives.

9. J. H. Worst, *Sixth Annual Report of the North Dakota Agricultural Experiment Station, Fargo, N.D., to the Governor, February 1, 1896* (Jamestown: Alert, 1896): 12; idem, *Tenth Annual Report of the North Dakota Agricultural Experiment Station, Agricultural College, N.D., to the Governor of North Dakota, February 1, 1900, Public Document No. 1* (Grand Forks: Herald, 1900): 10; and idem, *Eleventh Annual Report of the North Dakota Agricultural Experiment Station, Agricultural College, N.D., to the Governor of North Dakota, February 1, 1901* (Bismarck: Tribune, 1901): 10.
10. For Shepperd's career see Bill G. Reid, *Five for the Land and Its People* (Fargo: North Dakota Institute for Regional Studies, 1989), 103-15.
11. Letters from S. G. Warrington to J. H. Shepperd, August 29, 1899, box 2, folder 10; Eugene Disney to J. H. Shepperd, August 26, 1901, box 3, folder 2; Marion Arlion to John Worst, October 23, 1900, box 2, folder 3; Sisters of Mercy to Agricultural College, June 4, 1901, box 3, folder 5; John Milsted to J. H. Shepperd, March 1, 1900, box 2, folder 7; and William Mayo to Experiment Station, January 22, 1901, box 1, folder 7, John H. Shepperd Papers, North Dakota Institute for Regional Studies.
12. See Charles E. Rosenberg, "Science, Technology, and Economic Growth: The Case of the Agricultural Experiment Station Scientist, 1875-1914," *Agricultural History* 45 (January 1971): 1-20.
13. J. H. Worst, *Fifth Bi-Ennial Report of the Board of Trustees of the North Dakota Agricultural College at Fargo, N.D. to the Governor of North Dakota, November 1, 1900, Public Document No. 9* (Grand Forks: Herald, 1900): 11; and idem, *Seventeenth Annual Report of the North Dakota Agricultural Experiment Station, Agricultural College, North Dakota, to the Governor of North Dakota, 1907* (Fargo: Walker Bros. and Hardy, 1907): 14-15.
14. Remarks of E. E. Kaufman, *Third Biennial Report . . . Agricultural College . . . 1895-96*: 49.
15. *Fourth Biennial Report of the Commissioner of Agriculture and Labor to the Governor of North Dakota for the Two Years Ending June 30, 1896* (Bismarck: Tribune, 1896): 60-61; and *Eighth and Ninth Biennial Report of the Commissioner of Agriculture and Labor to the Governor of North Dakota for the Term Ending June 30, 1906* (Bismarck: Tribune, 1906): 25, 30, and 75-77.
16. L. R. Waldron, "Some Principles of Dry Farming," North Dakota Agricultural College Government Agricultural Experiment Station of North Dakota, *Bulletin No. 97* (Fargo: March 1912): 462. For the promotion of dairy farming at the NDAES see my article "The North Dakota Agricultural Experiment Station and the Struggle to Create a Dairy State," *Agricultural History* 63 (Spring 1989): 174-86.
17. C. B. Waldron, "Trees and Fruit in North Dakota," North Dakota Agricultural College Government Agricultural Experiment Station, *Bulletin No. 59*

(Fargo: March 1904): 358; idem, "Windbreaks and Hedges," North Dakota Agricultural College Government Agricultural Experiment Station for North Dakota, *Bulletin No.* 88 (Fargo: June 1910): 11; and idem, *Sixteenth Annual Report of the North Dakota Agricultural Experiment Station, Agricultural College, North Dakota, to the Governor of North Dakota, 1906* (Fargo: Walker Bros. and Hardy, 1906): 51.

18. J. H. Worst, forward to Wallace Manikowske, "Windmill Electric Lighting and Power," North Dakota Agricultural College Government Experiment Station, *Bulletin No. 105* (Fargo: August 1913): 7.
19. H. L. Bolley, "Flax Wilt and Flax Sick Soil," Government Agricultural Experiment Station for North Dakota, *Bulletin No. 50* (Fargo, October 1901); idem, "Flax and Flaxseed Selection," Government Agricultural Experiment Station for North Dakota, *Bulletin No. 55* (Fargo: March 1903); idem, "Flax Culture," North Dakota Agricultural College Government Agricultural Experiment Station for North Dakota, *Bulletin No. 71* (Fargo: October 1906); *Fourth Biennial Report of the Commissioner of Agriculture and Labor . . . 1896*: 40; and *Eighth and Ninth Biennial Report of the Commissioner of Agriculture and Labor . . . 1906*: 42.
20. Letter from Charles H. Thornton to H. L. Bolley, April 3, 1941, box 24, "Th" file, p. 1, H. L. Bolley Papers, North Dakota Institute for Regional Studies.
21. Letter from H. L. Bolley to George E. Miller, February 9, 1911, box 13, "Miller, G.E." file, p. 1, Bolley Papers, NDIRS; and H.L. Bolley, *Twenty-Second Annual Report of the North Dakota Agricultural Experiment Station, Agricultural College, North Dakota, to the Governor of North Dakota, 1912* (n.p., n.d.): 44 and 72.
22. *Eleventh Annual Report . . . North Dakota Agricultural Experiment Station. . . 1901*: 57.
23. Letter from J. H. Shepperd to M. A. Carleton, March 22, 1904, box 23, folder 1, letterpress p. 108; letters from O. A. Thompson to J. H. Shepperd, July 3 and 19, 1904, box 8, folder 4; and letter from E. F. Ladd to J. H. Shepperd, September 18, 1904, box 1, folder 2, p. 1, Shepperd Papers, NDIRS.
24. The Campbell System, enthusiasm for it in the West, scientific skepticism regarding it, and the dry farming movement in general are all covered in Mary W. M. Hargreaves, *Dry Farming in the Northern Great Plains, 1900-1925* (Cambridge: Harvard University Press, 1957).
25. J. H. Shepperd and Joseph A. Jeffery, "A Study of Methods of Cultivation. I. At the Station. II. At Other Points in the State," Government Agricultural Experiment Station for North Dakota, *Bulletin No. 29* (Fargo: September 1897), J. H. Shepperd and A. M. Ten Eyck, "Cultivation Experiments With Wheat and a Special Study of Moisture and Temperature of the Soils Under the Campbell and Ordinary Treatments," North Dakota Agricultural College Government Agricultural Experiment Station for North Dakota, *Bulletin No. 38* (Fargo, 1899); and J. H. Shepperd and A. M. Ten Eyck, "Wheat

Farming Experiments and Soil Moisture Studies," North Dakota Agricultural College Government Agricultural Experiment Station for North Dakota, *Bulletin No. 48* (Fargo: June 1901): 736. Shepperd's pride in exposing the flaws of the Campbell System is reflected in his letter to E. W. Allen, June 24, 1908, OES, North Dakota, Jan. 1902-June 1937, National Archives.

26. Letter from J. E. Phelan to John H. Worst, February 14, 1905, box 9, folder 7; and letter from J. H. Shepperd to J. E. Phelan, March 4, 1905, box 9, folder 8, p. 1, Shepperd Papers, NDIRS. For Station suggestions regarding farming in the West see L. R. Waldron, "Hints to Homesteaders," North Dakota Agricultural College Government Agricultural Experiment Station of North Dakota *Bulletin No. 74* (Fargo: January 1907); and L. R. Waldron, "Some Principles of Dry Farming," North Dakota Agricultural College Government Agricultural Experiment Station of North Dakota, *Bulletin No. 96* (Fargo: January 1912).
27. Remarks of J. H. Shepperd, *First Annual Conference of the Bankers Committees on Agricultural Development and Education, Minneapolis-St. Paul, October 18-19, 1911*: 70.
28. Letter from M. A. Carleton to J. H. Shepperd, February 18, 1903, box 6, folder 3, and letter from W. F. Miller to North Dakota Agricultural Experiment Station, September 26, 1904, box 9, folder 3, Shepperd Papers, NDIRS. The long and complex struggle over durum is detailed in Mary W. M. Hargreaves, "The Durum Wheat Controversy," *Agricultural History* 42 (July 1968): 211-29.
29. Untitled editorials, *Northwestern Miller* 56 (September 2 and 9, 1903): 507 and 563.
30. E. F. Ladd, *Twenty-Second Annual Report . . . North Dakota Agricultural Experiment Station . . . 1912*: 17; and letter from E. F. Ladd to A. C. True, April 21, 1900, OES, North Dakota, Oct. 1888-Dec. 1901, National Archives.
31. Ladd's judgments regarding durum prices are found in "1. Milling and Baking Test; 2. Tests of Flour Sold in North Dakota," North Dakota Agricultural College Government Agricultural Experiment Station for North Dakota, *Bulletin No. 82* (Fargo: October 1908): 722. The *Miller*'s attitude toward experiment station milling tests can be found in an untitled editorial, *Northwestern Miller* 62 (May 10, 1905): 327. See also E. F. Ladd and C.H. Bailey, "Wheat Investigations: Milling, Baking and Chemical Tests," North Dakota Agricultural College Government Agricultural Experiment Station for North Dakota, *Bulletin No. 89* (Fargo: September 1910), and Hargreaves, "The Durum Wheat Controversy."
32. E. F. Ladd and R. E. Stallings, "Bleaching of Flour," North Dakota Agricultural College Government Agricultural Experiment Station for North Dakota, *Bulletin No. 72* (Fargo: November 1906): 227; Editorial, "The Bleaching Process," *Northwestern Miller* 76 (October 28, 1908): 220; and Harry Snyder,

"Effect of Bleaching on Flour as Food," *Northwestern Miller* 76 (October 7 and 14, 1908): 23-24, 40, and 91-92. See also Hargreaves, "The Durum Wheat Controversy."

33. Editorial, "Bleaching Under Fire," *Northwestern Miller* 77 (January 20, 1909): 145; and "North Dakota Bleaching Decision," *Northwestern Miller* 77 (January 13, 1909): 85 and 98. The bleaching controversy and its resolution is detailed in Hargreaves, "The Durum Wheat Controversy."
34. Remarks of L. R. Waldron, *Fourteenth Annual Report of the North Dakota Agricultural Experiment Station to the Governor* (Bismarck: Tribune, 1904): 45.
35. E. F. Ladd, "Food Products and Their Adulteration," Government Agricultural Experiment Station for North Dakota, *Bulletin No. 53* (Fargo: September 1902): 149-50. Ladd's career as a pure-food advocate is chronicled by Ralph J. Kane, "Edwin Fremont Ladd: North Dakota's Pure-Food Crusader" (Master's thesis, University of North Dakota, 1960).
36. E. F. Ladd, "1. Foods and Food Products; 2. Whiskey and Other Beverages; 3. Drugs and Medicines," North Dakota Agricultural College Government Agricultural Experiment Station of North Dakota, *Bulletin No. 86* (Fargo: December 1909): 80.
37. E. F. Ladd and G. A. Abbott, "Some Ready Mixed Paints," North Dakota Agricultural College Government Agricultural Experiment Station of North Dakota, *Bulletin No. 86* (Fargo: December 1909): 80; E. F. Ladd and C. D. Holley, "Paints and Paint Products," Government Agricultural Experiment Station of North Dakota, *Bulletin No. 67* (Fargo: December 1905), and E. F. Ladd, "Practical Paint Tests in 1907," North Dakota Agricultural College Government Agricultural Experiment Station of North Dakota, *Bulletin No. 81* (Fargo: September 1908).
38. Robinson, *History of North Dakota*, 254. For the attitude of the OES regarding regulatory work see *Annual Reports of the Department of Agriculture for the Fiscal Year Ended June 30, 1901* (Washington,D.C.: Government Printing Office, 1901): 185. For the typically negative attitude of one station toward regulatory work see C. Clyde Jones, "An Agricultural College's Response to a Changing World," *Agricultural History* 42 (October 1968): 286-87.
39. Letter from J. H. Worst to E. W. Allen, March 24, 1903, OES, North Dakota, Jan. 1902-June 1937, National Archives.
40. The background and significance of the Adams Act is detailed in Charles E.Rosenberg, "The Adams Act: Politics and the Cause of Scientific Research,"*Agricultural History* 38 (January 1964): 3-12.
41. Remarks of J. H. Worst, *Twenty-Third Annual Report of the North Dakota Agricultural Experiment Station, Agricultural College, North Dakota, to the Governor of North Dakota, 1913* (n.p., n.d.): 6. For an especially good example of staff aggressiveness see the *Thirteenth Annual Report of the North Dakota Agricultural Experiment Station to the Governor of North Dakota* (Bismarck: Tribune, 1903).

42. For the development of agricultural science see Margaret W. Rossiter, "The Organization of the Agricultural Sciences," in Alexandra Oleson and John Voss, eds., *The Organization of Knowledge in Modern America, 1860-1920* (Baltimore: Johns Hopkins University Press, 1979), 211-48.

CHAPTER 4
THE GREAT CONTROVERSY

1. Editorial, "Now For a Showdown," *Fargo Daily Courier-News* 7 (August 29,1914): 4.
2. Letter from H. L. Bolley to C. W. Buttz, February 1, 1916, box 28, "Personal" file, H. L. Bolley Papers, North Dakota Institute for Regional Studies.
3. Raymond Pearl, "The Need for Endowed Agricultural Research," *Science* 37 (May 9, 1913): 708.
4. C. E. Thorne, "The Progress of Education and Research in Agriculture," *Proceedings of the Thirtieth Annual Convention of the Association of American Agricultural Colleges and Experiment Stations, Held At Washington, D.C., November 15-17, 1916* (Burlington, Vt.: Free Press, (1917): 51.
5. For a discussion of the emphasis on farm management see Gould P. Colman, *Education and Agriculture: A History of the New York State College of Agriculture at Cornell University* (Ithaca, New York: Cornell University Press, 1963), 225-26, 308, and 315; and Richard S. Kirkendall, "The Agricultural Colleges: Between Tradition and Modernization," *Agricultural History* 60 (Spring 1986): 12-14.
6. For case studies of the emerging orientation of agricultural research see Ronald L. Nye, "Federal vs. State Agricultural Research Policy: The Case of California's Tulare Experiment Station, 1888-1909," *Agricultural History* 57 (October 1983): 436-49, and Colman, *Education and Agriculture*, 239-40. See also my "Publicly-Sponsored Agricultural Research from An Historical Perspective," in Kenneth A. Dahlberg, ed., *New Directions for Agricultural Research: Neglected Dimensions and Emerging Alternatives* (Totowa, N.J.: Rowman and Allanheld, 1986), 107-31. For the problem of static productivity see my *Resisted Revolution: Urban America and the Industrialization of Agriculture, 1900-1930* (Ames: Iowa State University Press, 1979), esp. 36-42.
7. James Wilson, "Report of the Secretary of Agriculture," *Annual Reports of the Department of Agriculture for the Fiscal Year Ending June 30, 1900* (Washington, D.C.: Government Printing Office, 1900): LVI; and remarks of A. C. True, *Annual Reports of the Department of Agriculture for the Fiscal Year Ended June 30, 1901* (Washington, D.C.: Government Printing Office, 1901): 180.
8. Remarks of A. C. True, *Annual Reports of the Department of Agriculture for the*

Fiscal Year Ended June 30, 1905 (Washington, D.C.: Government Printing Office, 1905): 441; and E. W. Allen, "Administration of Experiment Station Work By Projects," *Proceedings of the Twenty-Eighth Annual Convention of the Association of American Agricultural Colleges and Experiment Stations, Held At Washington, D.C., November 11-13, 1914* (Montpelier, Vt.: Capital City Press, 1915): 234.

9. Remarks of A.C. True, *Annual Reports . . . Department of Agriculture . . . 1901*: 183.
10. Eugene Davenport, "The Outlook for Agricultural Science," *Science* 45 (February 16, 1917): 158.
11. T. E. Stoa, *The Agronomy Department—A History 1890-1960* (Fargo: North Dakota State University, 1966): 3.
12. Report from H. L. Bolley to J. H. Worst, December 31, 1913, box 9, "Reports, Various" file, Bolley Papers, NDIRS.
13. H. L. Bolley, "Retrenchment-Recommendations," ca. March 15, 1912, box 33, "Worst 1911-1912" file, Bolley Papers, NDIRS.
14. William C. Hunter raises the generational issue in *Beacon Across the Prairie: North Dakota's Land-Grant College* (Fargo: North Dakota Institute for Regional Studies, 1961), 43.
15. H. L. Bolley, "How a Farmer Can Tell Whether His Land Is Wheat Sick or Not," *Fargo Forum* (June 26, 1911): 12.
16. Letter from A. R. Rogers to H. L. Bolley, July 7, 1911, box 28, "Rogers, A. R.-*Special*" file; letter from H. L. Bolley to A. F. Woods, July 19, 1911, box 28, "Woods, A. F." file; letter from J. H. Worst to H. L. Bolley, July 20, 1911, box 33, "Worst-Specials" file; and letter from H.L. Bolley to J.H. Worst, July 23, 1911, box 33, "Worst-Specials" file, Bolley Papers, NDIRS.
17. H.L. Bolley, "Wheat: Soil Trouble and Seed Deterioration. Causes of Soil Sickness in Wheat Lands. Possible Methods of Control. Cropping Methods With Wheat," North Dakota Agricultural College Government Agricultural Experiment Station for North Dakota, *Bulletin No. 107* (Fargo: December 1913): 6; and Cyril Hopkins, "The Bread Supply," *Science* 38 (August 22, 1913): 479-81. See also H. L. Bolley, "Cereal Cropping: Sanitation, a New Basis for Crop Rotation, Manuring, Tillage and Seed Selection," *Science* 38 (August 22, 1913): 249-59.
18. *First Annual Conference of the Bankers Committees on Agricultural Development and Education, Minneapolis-St. Paul, October 18-19, 1911* (n.p., n.d.). My delineation of the development of the Better Farming Association is drawn from several sources, including the *Fargo Forum* and *Daily Courier-News*, the *Co-Operator's Herald*, the *Nonpartisan Leader*, the *North Dakota Banker*, the Bolley and Shepperd papers at the North Dakota Institute for Regional Studies, the Great Northern Papers at the Minnesota State Historical Society, Hunter's *Beacon Across the Prairie*, and an undated manuscript entitled "The Looting of the North Dakota Agricultural College. Some Inside

History," held by the North Dakota Institute for Regional Studies. I believe the author of this document to have been Worst, though it might have been written by Bolley or engineering professor E. S. Keene. Though these sources diverge dramatically on interpretation, they show a remarkable degree of agreement regarding the exact nature and timing of key events.

19. "War Breaks Out At A.C. Again Over Gang Control of the School," *The Co-Operator's Herald* 3 (February 25, 1916): 7-8.
20. "List of Subscribers and Amount Subscribed to the N.D. Better Farming Association," n.d., president's subject file no. 5344, Great Northern Papers, Minnesota State Historical Society.
21. Letter from P. L. Howe to S. Hegstad, January 8, 1913, president's subject file no. 4807, Great Northern Papers, MSHS; R. S. Adams, "President's Annual Address—Grand Forks Convention, June 17 1913," *North Dakota Banker* 1 (July 1913): 2; and letter from Carl R. Gray to W. C. Allen, October 8, 1912, Great Northern Papers, MSHS, president's subject file no. 4807.
22. Thomas P. Cooper, "Work of the Better Farming Association of North Dakota," *North Dakota Banker* 2 (October 1913): 13.
23. For the American Society of Equity see Elwyn B. Robinson, *History of North Dakota* (Lincoln: University of Nebraska, 1966): 276-78 and 330-31; A. E. Walley, "An Open Letter to the Fargo Forum," *The Co-Operator's Herald* 1 (April 10, 1914): 5.
24. For railroad development projects in North Dakota see Roy V. Scott, *Railroad Development Programs in the Twentieth Century* (Ames: Iowa State University Press, 1985); and Mary W. M. Hargreaves, *Dry Farming in the Northern Great Plains*, 1900-1925 (Cambridge: Harvard University Press, 1957).
25. See, for example, the letter from J. H. Shepperd to Thomas P. Cooper, July 1, 1913, box 12, folder 2, John H. Shepperd Papers, North Dakota Institute for Regional Studies.
26. Letter from J. H. Worst to W. A. Allen, May 10, 1913, and letter from J. H. Worst to A. C. True, April 22, 1913, Record Group 164, Office of Experiment Stations, State Correspondence, General (1888-1937) North Dakota, Jan. 1902-June 1937, National Archives.
27. Remarks of Thomas P. Cooper, *North Dakota Agricultural Experiment Station Twenty-Fifth Annual Report to the Governor of North Dakota* (February 1915): 4; and idem, *Annual Convention of the Association of American Agricultural Colleges and Experiment Stations . . . 1916*: 279.
28. *The Work of the Experiment Station, Report of Director Thomas P. Cooper* (Agricultural College, N.D., January 1916): 4 and 3.
29. Letter from Thomas Cooper to H. L. Bolley, January 28, 1914, box 32, "Cooper-Orders" file, Bolley Papers, NDIRS. Also in the Bolley Papers, see letter from Thomas Cooper to H. L. Bolley, February 4, 1914, box 32, "Cooper on Pure Seed Matters" file; letter from H. L. Bolley to Thomas Cooper, February 5, 1914, box 32, "Cooper-Orders" file; letter from H. L.

Bolley to Prof. _______, March 14, 1914, box 32, "Cooper on Projects and Budget" file; letter from Thomas Cooper to H. L. Bolley, April 19, 1915, box 3, "Budgets 1915-1920" file; and letter from H. B. Humphrey to H. L. Bolley, April 7, 1915, box 4, "Humphrey, H.B." file.

30. Letter from H. L. Bolley to M. L. Wilson, January 25, 1916, box 7, "Wilson, M. L." file, Bolley Papers, NDIRS. See also the letter from H. L. Bolley to A. F. Woods, April 14, 1914, box 28, "Personal" file, Bolley Papers, NDIRS.
31. George L. Nelson, "An Unwarranted Outrage," *Co-Operator's Herald* 1 (January 30, 1914): 1 and 3.
32. The ASE belabored Worst mercilessly over the the Fargo incident. See, for example, Author Unknown, "Worst a Reactionary," *Co-Operator's Herald* 1 (February 20, 1914): 5; and George L. Nelson, "Open Letter to President Worst," *Co-Operator's Herald* 1 (March 6, 1914): 4-5. Statement of H. D. Long, n.d., box 28, "Personal" file, Bolley Papers, NDIRS.
33. "Forum Presents Some Facts Regarding the Alleged A.C. Muddle," *Fargo Forum* (September 5, 1914): 1, and George L. Nelson, "N.D.A.C. Stand—and Oil Merger [sic]," *Co-Operator's Herald* 1 (September 11, 1914): 1.
34. George L. Nelson, "An Astounding Revelation," *Co-Operator's Herald* 1 (August 28, 1914): 1 and 8; and idem, "NDAC Board Holds Fake Quiz," *Co-Operator's Herald* 2 (October 2, 1914): 5.
35. Upton Sinclair's exposé of corporate control of higher education, *The Goose-Step: A Study of American Education* (Pasadena, Calif.: Upton Sinclair, 1923), prominently features the BFA/NDAES affair.
36. Letter from T. A. Hoverstad to J. H. Shepperd, n.d., box 13, folder 7, Shepperd Papers, NDIRS.
37. For a discussion of the general problem of social values and scientific research see Loren R. Graham, *Between Science and Values* (New York: Columbia University Press, 1981). In totalitarian societies this phenomenon is especially clear. See David Joravsky, *The Lysenko Affair* (Cambridge: Harvard University Press, 1970); and Alan D. Beyerchen, *Scientists Under Hitler: Politics and the Physics Community in the Third Reich* (New Haven: Yale University Press, 1977). For the influence of business and industry on the direction of scientific research during the progressive period see David F. Noble, *America By Design: Science, Technology, and the Rise of Corporate Capitalism* (New York: Knopf, 1979).
38. George L. Nelson, "Big Rumpus at the A.C. Arrives on Time as Herald Predicted," *Co-Operator's Herald* 2 (December 2, 1914): 1 and 8.
39. Letter from H. L. Bolley to L. F. Crawford, January 21, 1916, box 28, "Personal" file, and letter from H. L. Bolley to J. H. Worst, February 5, 1916, box 33, "Worst 1914-15" file, Bolley Papers, NDIRS. For Bolley's public comments see "Disease Not the Greatest," *Grand Forks Herald*, January 29, 1916, box 28, "Personal" file, Bolley Papers, NDIRS.
40. "Worst, Regents, Faculty, Students and Public Back of Ladd and Reorganiza-

tion of the Agricultural College," *Fargo Forum* (February 29, 1916): 5; "Meeting At Armory Saturday Resulted in Very Frank Discussion of the Situation," *Fargo Forum* (February 29, 1916): 1 and 5; and "President Worst Appears Before Board of Regents," *Fargo Daily Courier-News* 8 (February 19, 1916): 1.

41. Editorial, "College Will Go Ahead," *Fargo Forum* (February 29, 1916): 4; editorial, "A Lull in the Storm," *Co-Operator's Herald* 3 (March 10, 1916): 4; editorial, "The Agricultural College," *North Dakota Banker* 4 (March 1916): 13; letter from H. L. Bolley to S. M. Coulter, March 20, 1916, box 28, "Personal" file, Bolley Papers, NDIRS, and letter from L. S. Bottenfield to J. H. Shepperd, May 7, 1916, box 13, folder 2, Shepperd Papers, NDIRS. See also "Worst, Regents, Faculty, Students and Public Back of Ladd . . . ," *Forum*. Ladd told Upton Sinclair that the board urged him to fire Bolley when he was appointed and that it later attempted to force him to resign the presidency; see Sinclair's *Goose-Step*, 205.
42. Letter from W. H. Mercer to E. F. Ladd, June 1, 1916, "Hearing of 1916" box, unmarked file, Bolley Papers, NDIRS.
43. H. L. Bolley, "Tentative Answer to Charges Filed Against H. L. Bolley by Messrs. Milbrath and Mercer," "Hearing of 1916" box, unmarked file, Bolley Papers, NDIRS. Renowned flax researcher Harold Flor, who worked with Bolley during the thirties and forties, credited him with being a skilled observer but contended that his research methods were very careless. Telephone interview with Harold Flor, July 19, 1984.
44. Statement of Nevada Evans, April 8, 1916; and letter from Casper I. Nelson to H. L. Bolley, June 11, 1916, "Hearing of 1916" box, unmarked file, Bolley Papers, NDIRS.
45. Robert L. Morlan's *Political Prairie Fire* (Minneapolis: University of Minnesota Press, 1959) remains the best book on the Nonpartisan League. William Lemke, a key League figure, took an especial interest in Cooper and his opponents at the Station. See Edward C. Blackorby, *Prairie Rebel: The Public Life of William Lemke* (Lincoln: University of Nebraska Press, 1963). It is probably not coincidental that Ladd, a man with shrewd political instincts, appointed the investigating committee within a few days after the NPL primary victory.
46. Editorial, "Will Be Missed," *Fargo Forum* 40 (December 11, 1917): 4, and letter from H. L. Bolley to L. T. Minehart, February 4, 1918, box 5, "Nolan—F.D.C." file, Bolley Papers, NDIRS.
47. Remarks of E. W. Allen, *Annual Reports of the Department of Agriculture for the Year Ended June 30, 1920* (Washington, D.C.: Government Printing Office, 1921): 454.
48. P. F. Trowbridge, "Report of the Director, July, 1917 to June 30, 1919," Agricultural Experiment Station, *Bulletin 136* (March 1920): 3.
49. Interview with T. E. Stoa, July 10, 1984.

50. Letter from J. E. Eastgate to J.H. Shepperd, September 4, 1919, box 12, folder 5, Shepperd Papers, NDIRS. For the League's opinion of Ladd see the letter from Senator Frazier's aide, W. D. Skeels, to NPL leader John N. Hagen, December 19, 1924, box 1, John Hagen Papers, State Historical Society of North Dakota.

CHAPTER 5

COOPERISM WITHOUT COOPER

1. H. L. Bolley, "The Land Grant Colleges—Their Contribution," March 10, 1921, box 2, "Address Bolley-Land Grant" file, H. L. Bolley Papers, North Dakota Institute for Regional Studies, p. 5; and letter from Joel H. Greene to H. L. Bolley, n.d. [probably 1928], box 2, "Bolley, Letters to" file, Bolley Papers, NDIRS.
2. E. A. Barnett, "New Federal Appropriations for Experiment Stations," *Proceedings of the Thirty-Third Annual Convention of the Association of American Agricultural Colleges and Experiment Stations Held At Chicago, Illinois, November 12-14, 1919* (Burlington, Vt.: Free Press, 1920): 257.
3. For the agricultural depression of the twenties see James H. Shideler's classic *Farm Crisis, 1919-1923* (Berkeley: University of California Press, 1957).
4. Elwyn B. Robinson, *History of North Dakota* (Lincoln: University of Nebraska Press, 1966), 372; and R. Wayne Newton and Alva Benton, "Some Tax Problems of North Dakota Farmers," North Dakota Agricultural Experiment Station, *Bulletin 203* (Fargo: October 1926): 22.
5. Interview with Christian Jensen, July 11, 1984. See also "Perry Fox Trowbridge," box 1, "Historical" file, North Dakota Agricultural Experiment Station Director's Office Papers, North Dakota Institute for Regional Studies.
6. R. W. Thatcher, "Ideals in Agricultural Research," *School and Society* 11 (March 27, 1920): 361.
7. William C. Hunter, *Beacon Across the Prairie: North Dakota's Land-Grant College* (Fargo: North Dakota Institute for Regional Studies, 1961), 114.
8. Remarks of P. F. Trowbridge, *Report of the Director of the North Dakota Agricultural Experiment Station, July 1, 1919, to June 30, 1920* (Agricultural College, March 1921): 3; interview with Glenn Smith, September 24, 1984; remarks of P. F. Trowbridge, *Experiment Station Progress: Annual Report of the Director of the North Dakota Agricultural Experiment Station, 1920-1921* (Agricultural College: April 1922): 3; and remarks of P. F. Trowbridge, *Report of the North Dakota Agricultural Experiment Station* (Agricultural College: January 1926): 3.
9. Letter from H. L. Bolley to P. F. Trowbridge, March 28, 1922, box 7,

"Trowbridge" file, H. L. Bolley Papers, North Dakota Institute for Regional Studies. For a federal complaint about inappropriate expenditures at the Station see the letter from E. W. Allen to P. F. Trowbridge, October 9, 1923, Office of Experiment Stations, Record Group 164, State Correspondence, General (1888-1937) North Dakota, Jan. 1902-June 1937, National Archives.

10. Letter from P. F. Trowbridge to H. L. Bolley, December 21, 1921, Box 7, "Trowbridge" file, Bolley Papers, NDIRS; and remarks of P. F. Trowbridge, *Proceedings of the Thirty-Sixth Annual Convention of the Association of Land-Grant Colleges Held At Washington, D.C., November 21-23, 1922* (Burlington, Vt.: Free Press, 1923): 186.
11. Jensen interview.
12. Rex Willard and O. M. Fuller, "Type-of-Farming Areas in North Dakota," North Dakota Agricultural Experiment Station, *Bulletin 212* (Fargo: July 1927): 16. For the Northwest Farm Managers Association see Hiram M. Drache, *Plowshares to Printouts: Farm Management As Viewed Through 75 Years of the Northwest Farm Managers Association* (Danville, Ill.: Interstate, 1985).
13. Remarks of C. E. Mangels, *Report of the North Dakota Agricultural Experiment Station* (Agricultural College, February 1924): 39; Alva Benton, "Dockage in Wheat in North Dakota Foster County [sic] in Detail," North Dakota Agricultural Experiment Station, *Bulletin 172* (Agricultural College: February 1924): 12; and Rex Willard , "Cost of Producing Crops in North Dakota," North Dakota Agricultural Experiment Station, *Bulletin 199* (Agricultural College, July 1926): 4-5.
14. Alva Benton and Orville Fuller, "Great Lakes—St. Lawrence Deep Waterway: Its Value to North Dakota," North Dakota Agricultural Experiment Station, *Bulletin 204* (Fargo: January 1927).
15. Interview with T. E. Stoa, July 10, 1984; Jensen interview.
16. Letter from P. F. Trowbridge to H. L. Bolley, September 18, 1920, box 33, "Trowbridge, 1919-20" file, pp. 1-2; and letter from P. F. Trowbridge to H. L. Bolley, September 29, 1921, box 33, "Trowbridge 1920-21" file, Bolley Papers, NDIRS
17. Letter from John Lee Coulter to the Board of Administration, March 15, 1929, p. 21, letter from P. F. Trowbridge to J. L. Coulter, March 18, 1929, Box 10, "Coulter re Seed Commissioner" file, and letter from H. L. Bolley to J. H. Worst, February 4, 1930, box 25, "Wr" file, Bolley Papers, NDIRS. On responsibility for potato certification see the letter from H. L. Bolley to G. A. Stenehjem, January 4, 1928, box 23, "Stenehjem" file, Bolley Papers, NDIRS.
18. Interview with Rodney Hastings, December 23, 1987.
19. Charles H. Ruzicka, "Williston Substation Report for Years 1914 to 1920," North Dakota Agricultural Experiment Station, *Bulletin 158* (May 1922): 3.
20. *Seventeenth Biennial Report of the Commissioner of Agriculture and Labor to the*

Governor of North Dakota for the Term Ending June 30, 1922 (Grand Forks: Normanden Publishing, 1922): 14-33; *Compiled Agricultural Statistics of North Dakota For the Period Ending June 30, 1932* (n.p., n.d.): 5-24.

21. Harry G. Anderson and Alva H. Benton, "Cooperative Marketing of Dairy Products," North Dakota Agricultural Experiment Station, *Bulletin 238* (Fargo: September 1930): 25.
22. A. F. Yeager, "Vegetable Varieties for North Dakota" and "Fruit Culture in North Dakota," North Dakota Agricultural Experiment Station *Bulletin 187* and *Bulletin 188* (Agricultural College, February 1925), and Alva Benton, "Poultry and Egg Marketing in North Dakota," North Dakota Agricultural Experiment Station, *Bulletin 215* (Fargo: February 1928).
23. Telephone interview with Harold Flor, July 19, 1984; and interview with Glenn Smith, September 24, 1985. L. R. Waldron's life and contributions are examined in Bill G. Reid's *Five for the Land and Its People* (Fargo: North Dakota Institute for Regional Studies, 1989), 123-24.
24. L. R. Waldron, "Winter Wheat in North Dakota," North Dakota Agricultural Experiment Station, *Bulletin 151* (September 1921); and T.E. Stoa, "The Agronomy Department—A History 1890-1960" (Fargo: North Dakota State University, 1966): 13.
25. Smith interview; C. E. Mangels, T. E. Stoa, and William Guy, "Protein Survey of the North Dakota Wheat Crops of 1925 and 1926," North Dakota Agricultural Experiment Station, *Bulletin 208* (Fargo: March 1927): 3; L. R. Waldron, "Hybrid Selections of Marquis and Kota: Comparative Study With Regard to Disease Resistance[,] Yield and Baking Quality," North Dakota Agricultural Experiment Station, *Bulletin 200* (Agricultural College, September 1926): 59; and interview with David Ebeltoft, September 12, 1985.
26. Reid, *Five for the Land and Its People*, 127-29.
27. Stoa, "Agronomy Department," 16; and "Northwest Crop Improvement Association," box 3, "Northwest Crop Improvement Association" file, NDAES Director's Office Papers, NDIRS.
28. Stoa, "Agronomy Department," 17.
29. Raymond A. Pearson, "The Great Responsibility of Land-Grant Colleges," *Proceedings of the Thirty-Eighth Annual Convention of the Association of Land-Grant Colleges Held At Washington, D.C., November 12-14, 1924* (Burlington, Vt.: Free Press, 1925): 25, and E. D. Ball, "The Future of Agricultural Research," *Scientific Monthly* 22 (May 1926): 436.
30. E.W. Allen, "Some Features of the First Year Under the Purnell Act," *Proceedings of the Fortieth Annual Convention of the Association of Land-Grant Colleges and Universities Held At Washington, D.C., November 16-18, 1926* (Northhampton, Mass.: Metcalf Printing, 1927): 160.
31. R. W. Thatcher, "Research Work in Agriculture," *Scientific Monthly* 25 (October 1927): 339; and B. Youngblood, "Relation of the Physical and Biological to the Social Sciences in Agricultural Research," *Proceedings of the*

Forty-Third Annual Convention of the Association of Land-Grant Colleges and Universities Held at Chicago, Illinois, November 12-14, 1929 (Burlington, Vt.: Free Press, 1930): 184.

32. See, for example, Rex Willard and O. M. Fuller, "Type-of-Farming Areas in North Dakota," NDAES *Bulletin 212*; and Rex Willard, "Prices of North Dakota Farm Products," North Dakota Agricultural Experiment Station, *Bulletin 232* (Fargo: December 1929).
33. Alva H. Benton and M. F. Peightal, "Farmers' Elevators in North Dakota: Organization, Operating Methods, Incomes, and Costs," North Dakota Agricultural Experiment Station, *Bulletin 206* (Fargo: February 1927); and Alva H. Benton and Harold E. Seielstad, "Cooperative Marketing of Livestock in North Dakota," North Dakota Agricultural Experiment Station, *Bulletin 223* (Fargo: November 1928).
34. E. A. Willson, "Social Organizations and Agencies in North Dakota," North Dakota Agricultural Experiment Station, *Bulletin 221* (Fargo: February 1927): 78.
35. "Experiment Station Progress: Report for the Biennium July 1, 1925 to June 30, 1927," North Dakota Agricultural Experiment Station, *Bulletin 217* (Fargo: March 1928): 3-4; and "Experiment Station Progress: Report for the Biennium July 1, 1927 to June 30, 1929," North Dakota Agricultural Experiment Station, *Bulletin 233* (Fargo: March 1930): 2 and 134.

CHAPTER 6
HITTING BOTTOM

1. H. L. Walster, "Backgrounds of Economic Distress in the Great Plains," in John C. Hoyt, *Drought of 1936: With Discussion of the Significance of Drought in Relation to Climate*, United States Department of Interior, Geological Survey, *Water-Supply Paper 820* (Washington, D.C.: United States Government Printing Office, 1938), 53. The Dust Bowl phenomenon has been much studied lately, though most scholarly attention has been devoted to the Southern Plains. Three important treatments, from sharply divergent ideological perspectives, are Paul Bonnifield's *The Dust Bowl: Men, Dirt, and Depression* (Albuquerque: University of New Mexico Press, 1979), R. Douglas Hurt's *The Dust Bowl: An Agricultural and Social History* (Chicago: Nelson-Hall, 1981), and Donald Worster's *Dust Bowl: The Southern Plains In the 1930s* (New York: Oxford University Press, 1979).
2. North Dakota Crop and Livestock Reporting Service, *North Dakota Wheat* (Fargo: North Dakota Crop and Livestock Reporting Service): 6, 42, 68, and 76. My remarks regarding North Dakota in the thirties are informed largely

by Elwyn B. Robinson's *History of North Dakota* (Lincoln: University of Nebraska Press, 1966): 396-419. See also D. Jerome Tweton and Daniel F. Rylance, *The Years of Despair: North Dakota in the Depression* (Grand Forks, N.D.: Oxcart Press, 1973).

3. Letter from M. D. Johnson to H. L. Bolley, March 21, 1933, box 17, "Johnson and Jones" file, p. 1, H. L. Bolley Papers, North Dakota Institute for Regional Studies; and letter from Treadwell Twichell to J. H. Shepperd, April 10, 1933, box 12, folder 8, John H. Shepperd Papers, North Dakota Institute for Regional Studies.
4. Interview with T. E. Stoa, July 10, 1984.
5. For federal activities in agriculture during the thirties see Theodore Saloutos, *The American Farmer and the New Deal* (Ames: Iowa State University Press, 1982).
6. Arthur Hyde, *Report of the Secretary of Agriculture*, 1932 (Washington, D.C.: United States Government Printing Office, 1932): 22; Andrew Boss, "The Relation of State Research and Extension Agencies to the Readjustment of Agriculture in the Evolving Economic Situation," *Proceedings of the Forty-Fourth Annual Convention of the Association of Land-Grant Colleges and Universities Held At Washington, D.C., November 17-19, 1930* (Burlington, Vt.: Free Press, 1931): 146; and T.O. Walton, "Relation of Land-Grant Colleges to the Present Situation," *Proceedings of the Forty-Eighth Annual Convention of the Association of Land-Grant Colleges and Universities Held At Washington, D.C., November 19-24, 1934* (Wilmington, Del.: Cann Brothers, 1935): 36-37.
7. C. B. Hutchison, "The Influence of Agricultural Research on Our Social and Economic Order," *Proceedings of the Forty-Fifth Annual Convention of the Association of Land-Grant Colleges and Universities Held At Chicago, Illinois, November 16-18, 1931* (Burlington, Vt.: Free Press, 1932): 229; and Raymond A. Pearson, "Science in Food Production," *Scientific Monthly* 35 (November 1932): 453-54.
8. Walton, "Relation of Land-Grant Colleges to the Present Situation," 37; and A. F. Woods, "Adjustments in Research to Meet the Present Emergency," *Proceedings of the Forty-Sixth Annual Convention of the Association of Land-Grant Colleges and Universities Held At Washington, D.C., November 14-16, 1932* (Burlington, Vt.: Free Press, 1933): 170.
9. Henry A. Wallace, "Research and Adjustment March Together," USDA Agricultural Adjustment Administration (Washington, D.C.: Government Printing Office, 1934): 2; and *Report of the Secretary of Agriculture, 1933* (Washington: United States Government Printing Office, 1933): 27. For science in the USDA under Wallace see Carroll W. Pursell, Jr., "The Administration of Science in the Department of Agriculture, 1933-1940," *Agricultural History* 42 (July 1968): 231-40.
10. M.L. Wilson, "The Roles of the United States Department of Agriculture

and the State Experiment Stations in Regional Research Programs," *Proceedings of the Forty-Ninth Annual Convention of the Association of Land-Grant Colleges and Universities Held At Washington, D.C., November 18-20, 1935* (Wilmington, Del.: Cann Brothers, 1935): 173.

11. Henry A. Wallace, "The Social Advantages and Disadvantages of the Engineering-Scientific Approach to Civilization," *Science* 79 (January 5, 1934): 3 and 4.
12. Interview with John Doubly, August 29, 1985. My account of events at the College and the Station during the thirties draws heavily on William C. Hunter's *Beacon Across the Prairie: North Dakota's Land-Grant College* (Fargo: North Dakota Institute for Regional Studies, 1961), 122-62.
13. Interview with Christian Jensen, July 11, 1984.
14. Interview with Glenn Smith, September 24, 1985.
15. The finances of the Station are detailed in "Experiment Station Progress for the Four-Year Period July 1, 1931 to June 30, 1935," North Dakota Agricultural Experiment Station, *Bulletin 286* (Fargo: November 1935): 96.
16. For complaints from the USDA about inappropriate use of Station funds and the state's failure to meet Federal matching requirements see the letter from James T. Jardine to H. C. Hanson, May 4, 1939, and the letter from B. Youngblood to H. L. Walster, April 5, 1940, box 1, "Office of Experiment Stations" file, North Dakota Agricultural Experiment Station, Director's Office Papers, North Dakota Institute for Regional Studies.
17. "Experiment Station Progress . . . 1931 to . . . 1935," NDAES, *Bulletin 286*: 96.
18. Letter from Cap Miller to Herbert C. Hanson, May 17, 1938, box 1, "Needs of Departments" file, p. 2, NDAES Director's Office Papers, NDIRS; Smith interview; interview Harold Flor, July 19, 1984; letter from H. L. Bolley to T. D. Hinebauch, October 18, 1933, box 16, "Him-Hiz" file, p. 1, Bolley Papers, NDIRS; letter from H. L. Bolley to Daniel J. Glomset, January 3, 1933, box 14, "Gi-Gl" file, Bolley Papers, NDIRS; and letter from J. H. Worst to J. H. Shepperd, June 25, 1933, Box 12, Folder 8, p. 2, Shepperd Papers, NDIRS.
19. "Experiment Station Progress . . . 1931 to . . . 1935," NDAES, *Bulletin 286*: 3-6.
20. Jensen interview; and letter from A. F. Yeager to William Langer, November 22, 1933, box 26, A-19 folder 8: "Agriculture College 1933," p. 2, William Langer Papers, Libby Library, University of North Dakota.
21. "Experiment Station Progress . . . 1931 to . . . 1935," NDAES *Bulletin 286*: 3, 7, 10, and 84. For the Station's interpretation of the Purnell Act see the letter from H. L. Walster to L. B. Bauer, July 12, 1946, Series 1193, Box 3, "Land Grant College Association" file, Agricultural Experiment Station Files, 1947-1949, State Archives, State Historical Society of North Dakota.
22. H. L. Walster, "Research and a North Dakota 'Live-At-Home' Program," North Dakota Agricultural Experiment Station, *Bimonthly Bulletin* 3 (Sep-

tember 1940): 2; P. F. Trowbridge and Albert Severson, "Pork: Slaughtering, Cutting and Curing," "Beef: Slaughtering, Cutting and Curing," and "Lamb: Slaughtering, Cutting and Curing," North Dakota Agricultural Experiment Station, *Circular 47*, *Circular 48*, *Circular 49* (Fargo: June 1933); O.A. Barton, "Killing and Dressing Poultry," North Dakota Agricultural Experiment Station, *Circular 50* (Fargo: June 1933); Esther Latzke, "Cooking and Canning Meats," North Dakota Agricultural Experiment Station, *Circular 51* (Fargo: January 1934); Howard F. McColly, "North Dakota Agricultural Experiment Station: Some Historical Memorabilia" (unpublished manuscript in the author's possession): 3-6; and interview with William Promersberger, December 28, 1987.

23. "Experiment Station Progress . . . 1931 to . . . 1935," NDAES, *Bulletin 286*: 31 and 63; and McColly, "North Dakota Agricultural Experiment Station": 7-8.
24. E. A. Willson, "Incomes and Cost of Living of Farm Families in North Dakota, 1923-1931," North Dakota Agricultural Experiment Station, *Bulletin 271* (Fargo: June 1933); Cap E. Miller and Willard O. Brown, "Farm Tenancy and Rental Contracts in North Dakota," North Dakota Agricultural Experiment Station, *Bulletin 289* (Fargo: November 1937); Donald G. Hay, "Changes in North Dakota Farm Population During 1936 and 1937," North Dakota Agricultural Experiment Station, *Bimonthly Bulletin* 1 (September 1938): 18.
25. H. R. Danielson and R. S. Kifer, "Farm Reorganization and Management in East Central North Dakota," North Dakota Agricultural Experiment Station, *Bulletin 273* (Fargo: June 1933), "Experiment Station Progress . . . 1931 to . . . 1935," NDAES, *Bulletin 286*: 48; and Walster, "Backgrounds of Economic Distress in the Great Plains."
26. T. E. Stoa, "The Agronomy Department—A History 1890-1960" (Fargo: North Dakota State University, 1966): 19.
27. Unfortunately, there is no complete scholarly study of William Langer's gubernatorial career. He is a major figure in Robinson's *History of North Dakota* and is also treated by Glenn H. Smith in "William Langer and the Art of Personal Politics," in Thomas W. Howard, ed., *The North Dakota Political Tradition* (Ames: Iowa State University Press, 1981), 123-50. While Smith focuses on Langer's senatorial career, he includes a number of valuable insights regarding Langer's character, personality, and political style. Glenn H. Smith is unrelated to veteran Station wheat breeder Glenn Smith.
28. See Hunter, *Beacon Across the Prairie*: 144-61.
29. Interview with Christian Jensen, August 27, 1985.
30. Numerous affidavits, box 26, A-19 folder 9: "Agricultural College, January 1, 1934—June 24, 1934," Langer Papers, Libby Library, UND. See also the letter from C. S. McCulloch to William Langer, November 25, 1933, box 26, A-19 folder 8: "Agricultural College, 1933," Langer Papers.
31. Letter from Kenneth McGregor to William Langer, August 7, 1937, box 85,

A-19 Folder 5: "Agriculture College Investigation and Removal, 1937-1938," Langer Papers, Libby Library, UND. See also the telegram from William Langer to Helen Stokke, August 5, 1937, box 81, A-19 folder 15: "Political File, June 1, 1937-Aug. 31, 1937," Langer Papers; letter from William Langer to F. W. Maier, September 24, 1937, box 85, A-19 folder 5: "Agriculture College Investigation and Removal, 1937-1938," Langer Papers; and "Fargo Pastor Scores Ruthless Dismissals From AC Staff in Sermon From His Pulpit," *Fargo Forum* 59 (August 2, 1937): 6.

32. See the testimony of John Hagan in "In the Matter of the Removal From Office of Jennie Ulsrud and R. A. Kinzer, as Members of the Board of Administration of the State of North Dakota," series 913, box 1, pp. 124 and 160, Secretary of State: Records of Investigations, State Archives, State Historical Society of North Dakota. See also "Hagan's Speech—1938," box 1, John N. Hagan Papers, State Historical Society of North Dakota; this was a standard campaign speech delivered by Hagan during his 1938 gubernatorial campaign. For other attacks on the College and the Station see an unsigned, undated letter to the Board of Administration and the letter from Theodore E. Wee to William Langer, August 25, 1937, box 85, A-19 folder 5: "Agricultural College Investigation and Removal, 1937-1938," Langer Papers, Libby Library, UND.
33. Smith interview; interview with Christian Jensen, August 27, 1985.
34. Jensen interview, August 27, 1985. For the political philosophy prevalent among early ecologists see Ronald C. Tobey, *Saving the Prairies: The Life Cycle of the Founding School of American Plant Ecology, 1895-1955* (Berkeley: University of California Press, 1981). The relationship between academic disciplines and political attitudes in a fascinating subject. For an insightful recent treatment see Lawrence Busch and William B. Lacy, *Science, Agriculture, and the Politics of Research* (Boulder, Colo.: Westview Press, 1983).
35. Hunter, *Beacon Across the Prairie*, 155-78.
36. Interview with Stanley Moore, June 18, 1987.
37. For a sample of Hanson's activities see the letter from Herbert C. Hanson to William Langer, August 17, 1937, box 3, "Governor of North Dakota" file, and "Staff Meeting—December 6, 1937," box 2, "Staff Meetings" file, NDAES Director's Office Papers, NDIRS. See also R. M. Harris and T. Sanderson, "Milling and Baking Quality of Wheat Produced in Years of Severe Rust in North Dakota," North Dakota Agricultural Experiment Station, *Circular 61* (Fargo: November 1937).
38. Herbert C. Hanson, "Importance of Research," North Dakota Agricultural Experiment Station, *Bimonthly Bulletin* 1 (November 1938): 2.
39. For Walster's efforts on behalf of the NDFU see the letter from F. D. Fromme to H. L. Walster, October 28, 1941, series 1193, box 1, "Office of Experiment Stations" file, Agricultural Experiment Station Administrative Files, 1938-1942, State Archives, State Historical Society of North Dakota.

40. Letter from H. L. Walster to W. A. Plath, November 17, 1942, box 1, "North Dakota Farm Bureau" file, AES Administrative Files, 1938-1942, State Archives; and letters from H. L. Walster to Charles D. Egley and Richard C. Joyce, September 21, 1940 and November 7, 1941, box 1, "Farmers Union Coop" file, AES Administrative Files, 1938-1942, State Archives.
41. Stoa, "The Agronomy Department," 19.
42. Letter from H. L. Walster to R. W. Trullinger, January 8, 1940, box 2, "Office of Experiment Stations" file, pp. 1-2, AES Administrative Files, 1938-1942, State Archives.
43. H. L. Walster, "Research Projects for Which There is Pressing Need in North Dakota and for Which Present State and Federal Funds are Inadequate," box 2, "Office of Experiment Stations" file, AES Administrative Files, 1938-1942, State Archives.
44. H. L. Walster, "A Statement of Policy With Respect to Personnel Submitted By the Director of the North Dakota Agricultural Experiment Station," box 1, "President and State Board NDAC" file, p. 2, AES Administrative Files, 1938-1942, State Archives.

CHAPTER 7

PLANNING FOR THE FUTURE AND THE PAST

1. H. L. Walster, "Agricultural Progress Through Research: Annual Report of the North Dakota Agricultural Experiment Station," North Dakota Agricultural Experiment Station, *Bulletin 371* (Fargo: January 1952): 4.
2. Baldur H. Kristjanson and C. J. Heltemes, "Handbook of Facts About North Dakota Agriculture," North Dakota Agricultural Experiment Station, Bulletin 357 (Fargo: June 1950): 7.
3. E. C. Auchter, "Agricultural Science and the People's Welfare," *Science* 96 (September 25, 1942): 284.
4. T. Swann Harding, *Two Blades of Grass: A History of Scientific Development in the U.S. Department of Agriculture* (Norman: University of Oklahoma Press, 1947), 326; and A. Richard Crabb, *The Hybrid-Corn Makers: Prophets of Plenty* (New Brunswick, N.J.: Rutgers University, 1947), 12.
5. Edmund E. Day, "Science and Social Progress," *Proceedings of the Association of Land-Grant Colleges and Universities, Fifty-Fifth Annual Convention, Chicago, Illinois, November 10-12, 1941* (New Haven, Conn.: Quinnipiack Press, 1942): 54.
6. See, for example, Frederick G. Brownell, "Super Cows and Chickens," *Ameri-*

can Magazine 141 (June 1946): 108, 110-11; Vance Johnson, "Farmer's New Bag of Tricks," *Science Digest* 21 (January 1947): 24-29; Georg Mann, "Farmer's Atomic Hired Man," *Science Digest* 26 (December 1949): 82-87; and "What's Ahead in Crops?" *Farm Journal* 78 (December 1954): 23.

7. Norwood Allen Kerr, *The Legacy: A Centennial History of the State Agricultural Experiment Stations, 1887-1987* (Columbia: Missouri Agricultural Experiment Station, 1987): 95. See also H. B. Tukey, "Research in Fundamental Biology and in Agriculture," *Science* 105 (May 30, 1947): 565-67; and Margaret W. Rossiter, "Graduate Work in the Agricultural Sciences, 1900-1970," *Agricultural History* 60 (Spring 1986): 37-57.
8. For examples of the thinking of two eloquent critics of the direction of agricultural change see Ralph Borsodi, "The Case Against Farming As a Big Business," *The Land* 6 (Winter 1947-48): 446-51; and A. Whitney Griswold, *Farming and Democracy* (New York: Harcourt, Brace, 1948). For the position of critical social scientists see Kerr, *The Legacy*, 136-37. A controversial study that provided a focal point in the shift in the rural social sciences was Walter Goldschmidt's *As You Sow* (New York: Harcourt, Brace, 1947).
9. For some examples of the optimistic mood see Richard K. Winslow, "Food: Not Less, But More," *Newsweek* 45 (April 25, 1955): 110, 112, and 114-15; Claude W. Gifford, "How to Get Bigger Markets for What We Have to Sell," *Farm Journal* 81 (May 1957): 29 and 152; and, "Cow-to-Consumer Research," *Country Gentleman* 124 (June 1954): 36.
10. For changes in rural North Dakota during and after World War II, see especially Elwyn B. Robinson, *History of North Dakota* (Lincoln: University of Nebraska Press, 1966): 424-27.
11. Donald B. Hay, James P. Greenlaw, and Lawrence E. Boyle, "Problems of Rural Youth in Selected Areas of North Dakota," North Dakota Agricultural Experiment Station, *Bulletin 293* (Fargo: June 1940): 5 and 63; James P. Greenlaw, Andrew J. Kovash, and F. B. Daniel, "Farm Population Continues to Decrease," North Dakota Agricultural Experiment Station, *Bimonthly Bulletin* 4 (July 1942): 6; James P. Greenlaw and Raymond Novak, "Fewer Farm People Again in 1940," North Dakota Agricultural Experiment Station, *Bimonthly Bulletin* 4 (July 1942): 11; and Morris H. Taylor, "Tax Delinquency As Related to Local Government Organization and Finance," North Dakota Agricultural Experiment Station, *Bimonthly Bulletin* 3 (March 1941): 27-30.
12. Robert L. Berger, "Land Values and Transfers, Second Quarter 1943 North Dakota," North Dakota Agricultural Experiment Station, *Bimonthly Bulletin* 6 (September 1943): 13; and Kristjanson and Heltemes, "Handbook of Facts About North Dakota Agriculture," 15.
13. Letter from H. L. Walster to John Moses, February 11, 1943, series 1193, box 1, "Farm Mobilization Committee (North Dakota)" file, Agricultural Experiment Station Administrative Files, 1938-1942, State Archives, State Historical Society of North Dakota.

14. Letter from H. L. Walster to R. W. Trullinger, July 9, 1942, series 1193, box 1, "Office of Experiment Stations" file, AES Administrative Files, 1938-1942, State Archives.
15. Letter from H. L. Walster to James T. Jardine, February 12, 1945, series 1193, box 3, "Office of Experiment Stations" file, AES Administrative Files, 1947-1949, State Archives. For Mida see L. R. Waldron, R. H. Harris, T. E. Stoa, and L. H. Sibbit, "Mida Wheat," North Dakota Agricultural Experiment Station, *Circular* 68 (Fargo: March 1944).
16. Interview with Thomas J. Conlon, July 1, 1987; E. C. Stakman, "Science in the Service of Agriculture," *Scientific Monthly* 60 (February 1949): 81; and interview with Al Kenner, June 17, 1987. Unless otherwise noted, my understanding of the 15B outbreak and the NDAES response to it is drawn largely from T. E. Stoa, "The Agronomy Department—A History 1890-1960" (Fargo: North Dakota State University, 1966).
17. Remarks of Glenn Smith and L. R. Waldron, "North Dakota's Agricultural Progress Through Research: Annual Report of the North Dakota Agricultural Experiment Station," North Dakota Agricultural Experiment Station, *Bulletin 365* (Fargo: January 1951): 28-29, and Stoa, "Agronomy Department," 28.
18. Interview with David Ebeltoft, September 12, 1985; and Stoa, "Agronomy Department," 32. See also "North Dakota's Agricultural Progress Through Research": NDAES, *Bulletin 371: 16.* For the relationship between the NDAES and the North Dakota Crop Improvement Association see Lars A. Jensen "Extension Agronomy History" (unpublished manuscript in the author's possession).
19. Interview with Glenn Smith, September 24, 1985.
20. Letter from J. W. Haw to Fred Aandahl, August 24, 1950, series 1193, box 4, "Stem Rust Damage and Meetings" file, AES Administrative Files, 1950-1951; letter from Milton R. Young to Philip V. Cardon, September 11, 1950, series 1193, box 4, "Stem Rust Damage and Meetings" file, AES Administrative Files, 1950-1951; letter from Donald G. Fletcher to G. C. Holm, January 14, 1954, series 1193, box 6, "Rust Prevention Association" file, AES Administrative Files, 1954-1956; letter from Donald G. Fletcher to __________, February 11, 1955, series 1193, box 7, "Rust Prevention Association" file, AES Administrative Files, 1954-1956; and letter from Glenn J. Talbott to Arlon G. Hazen, January 3, 1953, series 1193, box 5, "Farmers Union" file, AES Administrative Files, 1952-1957, State Archives.
21. Interview with Wayne Owens, August 29, 1985; and interview with Bryce Streibel, June 17, 1987.
22. Rainer Schickele, "Variability of North Dakota Farm Income," North Dakota Agricultural Experiment Station, *Bimonthly Bulletin* 11 (July-August 1949): 207-8. See also Rainer Schickele and Reuben Engelking, "Land Values and the Land Market in North Dakota," North Dakota Agricultural

Experiment Station, *Bulletin 353* (Fargo: June 1949); Rainer W. Schickele, "Variable Payment Plans: New Frontier in Farm Credit Arrangements," North Dakota Agricultural Experiment Station, *Bimonthly Bulletin* 15 (September-October 1952): 33-36; Philip J. Thair, "Stabilizing Farm Income Against Crop Yield Fluctuations," North Dakota Agricultural Experiment Station, *Bulletin 362* (Fargo: September 1950); and idem, "Federal Crop Insurance: Should You Take It Out in 1952?" North Dakota Agricultural Experiment Station, *Bimonthly Bulletin* 14 (January-February 1952): 90-97.

23. See, for example, Thomas J. Conlon, Raymond J. Douglas, and Leroy Moomaw, "Rotation and Tillage Investigations at the Dickinson Experiment Station, Dickinson, North Dakota," North Dakota Agricultural Experiment Station, *Bulletin 383* (Fargo: May 1953).
24. See George F. Will, "Tree Ring Studies in North Dakota," North Dakota Agricultural Experiment Station, *Bulletin 338* (Fargo: April 1946).
25. For two examples of irrigation research at the NDAES see Arlon G. Hazen, "Irrigation Trials in Western North Dakota," North Dakota Agricultural Experiment Station, *Bulletin 358* (Fargo: March 1950); and Rex Helfinstine and L. W. Schaffner, "Irrigation and Dryland Farming Can Work Together on the Cannonball River," North Dakota Agricultural Experiment Station, *Bulletin 385* (Fargo: June 1953). The Garrison Diversion Project and the position of the NDAES in regard to it is fully explained in Paul Edward Kelly, "Under the Ditch: Irrigation and Garrison Diversion" (Master's Thesis: North Dakota State University, 1989).
26. H. L. Walster, "Moving Ahead in Farm Research," in "North Dakota's Agricultural Progress Through Research: Annual Report of the North Dakota Agricultural Experiment Station," North Dakota Agricultural Experiment Station *Bulletin 356* (Fargo: January 1950): 5.
27. C. Jensen, L. D. Beck, and Emily Pluth, "Standardization of North Dakota Creamery Butter," North Dakota Agricultural Experiment Station, *Bimonthly Bulletin* 11 (September-October 1948): 7-10; L. A. Fourt, "Butterfat Marketing: How Does North Dakota Shape Up?" North Dakota Agricultural Experiment Station *Bimonthly Bulletin* 14 (January-February 1952): 75-81; L. W. Schaffner and O. E. Jacobson, "Production Ups and Downs Hinder Quality Egg Marketing Program," North Dakota Agricultural Experiment Station, *Bimonthly Bulletin* 16 (September-October 1953): 3-5; L. A. Fourt and G. A. Kristjanson, "North Dakota's Dairy Marketing Problems in Historical Perspective," North Dakota Agricultural Experiment Station, *Bulletin 393* (Fargo: June 1954); and L. W. Schaffner, "Marketing Table-Stock Potatoes," North Dakota Agricultural Experiment Station, *Bimonthly Bulletin* 10 (January-February 1948): 116-18.
28. Remarks of Roy Gilcreast in "North Dakota's Agricultural Progress Through Research," NDAES, *Bulletin 371*, 63; and Roy M. Gilcreast, "Electricity—How Much? What For? On Farms in North-Central North Dakota," North

Dakota Agricultural Experiment Station, *Bulletin 379* (Fargo: October 1952): 26-27.

29. H. L. Walster, "The Agricultural Highway," North Dakota Agricultural Experiment Station, *Bimonthly Bulletin* 3 (January 1941): 2; and Rainer Schickele, "Role of the Family Farm in Modern Society," North Dakota Agricultural Experiment Station, *Bimonthly Bulletin* 10 (July-August 1948): 209.
30. Remarks of C. B. Haver and Reuben Engelking, p. 71, and remarks of Rainer Schickele and A.H. Anderson, p. 67, "North Dakota's Agricultural Progress Through Research," NDAES, *Bulletin 371*.
31. Baldur H. Kristjanson, "What About Our Large Farms in North Dakota?"North Dakota Agricultural Experiment Station, *Bulletin 360* (Fargo: July 1950); Baldur H. Kristjanson and Gordon A. Johnsgard, "Toward Scientific Farm Property Assessment," North Dakota Agricultural Experiment Station, *Bimonthly Bulletin* 13 (November-December 1950): 47-49; and Baldur H. Kristjanson and Jacob A. Brown, "Credit Needs of Beginning Farmers in Selected Areas of North Dakota," North Dakota Agricultural Experiment Station, *Bulletin 386* (June 1953).
32. This fourth and final purge at the College is outlined in William C. Hunter's *Beacon Across the Prairie: North Dakota's Land-Grant College* (Fargo: North Dakota Institute for Regional Studies, 1961), 203-10.
33. Interview with Joseph Zubriski, July 21, 1987.
34. Interview with John Callenbach, September 24, 1984; and interview with Richard L. Post, August 18, 1987. For an overview of the social and environmental dimensions of chemical control of insects see John H. Perkins, *Insects, Experts, and the Insecticide Crisis: The Quest for New Pest Management Strategies* (New York: Plenum, 1982).
35. Callenbach interview.
36. Post interview.
37. J. A. Munro, "Drafting Insect 'Soldiers' to War on Plant Parasites," North Dakota Agricultural Experiment Station, *Bimonthly Bulletin* 13 (July-August 1951): 231. Also, Post interview.
38. Post interview; interview with William Dinusson, July 23, 1987, and interview with Jack Carter, October 1, 1987. The Hultz building program is detailed in Hunter, *Beacon Across the Prairie*, 197-202.
39. Smith interview.
40. Ibid.
41. Letter from Nolan F. Farris to H. L. Walster, June 13, 1951, series 1193, box 4, "Office of Experiment Stations Annual Inspection" file, p. 1, AES Administrative Files, 1950-1951; and letter from Ralph B. Nestler to H. L. Walster, July 3, 1952, series 1193, box 5, "Office of Experiment Stations" file, AES Administrative Files, 1952-1957, State Archives.
42. Letter from H. L. Walster to R. W. Trullinger, July 1, 1952, Series 1193, Box

5, "Office of Experiment Stations" File, p. 1; and letter from H. L. Walster to Lulu Evanson, February 1, 1952, series 1193, box 5, "Farmers Union" file, AES Administrative Files, 1952-1957, State Archives.

43. H. L. Walster, "The Land Grant College and Agricultural Research," North Dakota Agricultural Experiment Station, *Bimonthly Bulletin* 15 (May-June 1953): 187.

CHAPTER 8

THE GOLDEN AGE

1. A. G. Hazen, "From the Director," *North Dakota Farm Research* 25 (January-February 1968): 16; "Show and Tell," *North Dakota Farm Research* 23 (September-October 1963): 3.
2. *U.S. Census of Agriculture: 1959: General Report-Vol. II-Statistics by Subjects* (Washington, D.C.: Government Printing Office, 1962): 975; and *1974 Census of Agriculture: Volume II-Statistics By Subject: Part 1-General Information: Procedures for Collection, Processing and Classification* (Washington,D.C.: Government Printing Office, 1978): 23.
3. *North Dakota Crop and Livestock Statistics* (Fargo: North Dakota Crop and Livestock Reporting Service, May 1979): 75; and *North Dakota Agricultural Statistics 1980* (Fargo: North Dakota Crop and Livestock Reporting Service, 1980): 8.
4. J. F. Carter, "Agronomy," *North Dakota Farm Research* 36 (November-December 1978): 9.
5. Interview with Peder Nystuen, July 22, 1987.
6. Interview with Leroy Schaffner, October 8, 1985; and interview with David Ebeltoft, September 12, 1985.
7. Interview with Harold Klosterman, August 19, 1987; and interview with Tom Conlon, July 1, 1987.
8. Interview with Glenn Smith, September 24, 1985; and interview with John Callenbach, September 24, 1985.
9. Interview with Stanley Moore, June 18, 1987; and interview with Arthur Link, June 30, 1987.
10. Interview with Al Kenner, June 17, 1987.
11. "Advisory Committees in Agriculture: North Dakota State University,"series 1193, box 19, "Vice President, Agriculture" file, p. 1, Agricultural Experiment Station Administrative Files, 1970-1971, State Archives, State Historical Society of North Dakota.
12. Interview with Richard Kiesling, August 17, 1987; and interview with Robert Johansen, August 17, 1987.

13. Klosterman interview.
14. Interview with George Sinner, July 2, 1987.
15. Letter from Arlon G. Hazen to R. M. Stout, Jr., June 23, 1975, box 23, "GNDA Greater North Dakota Association" file, AES Administrative Files, State Archives.
16. Letter from Arlon G. Hazen to Milton R. Young, February 1, 1966, box 15, "U.S. Congressmen" file, p. 1; and letter from Milton R. Young to Arlon Hazen, February 1, 1971, box 18, "Misc. State Offices and Officials (incl. U.S. Congressman [*sic*] from N.D.)" file, AES Administrative Files, State Archives.
17. Sinner interview.
18. Interview with Lois Jones, June 16, 1987; and interview with Keith Bjerke,June 16, 1987.
19. Interview with William Guy, July 2, 1987; and interview with Kelly Miller,June 17, 1987.
20. Interview with Clark Jenkins, July 22, 1987; and interview with Jack Dahl, July 2, 1987.
21. Arlon G. Hazen, "Summary of Remarks in Addressing North Dakota County Extension Agents at NDAC, August 16, 1957," box 9, "Director, Manuscripts and Speeches" file, p. 2; and letter from Arlon G. Hazen to George K. Brinegar, April 13, 1965, box 14, "Assn. of State University and Land Grant Colleges" file, p. 1, AES Administrative Files, State Archives.
22. Letter from Arlon G. Hazen to Paul Posel, February 23, 1977, Box 25, "Inquiries Answered" file, p. 2, AES Administrative Files, State Archives.
23. Letter from K. A. Gilles to Kenneth E. Raschke, March 30, 1972, box 20, "Vice President for Agriculture K.A. Gilles (folder 2)" file, AES Administrative Files, State Archives.
24. Carter, "Agronomy," 9. See also Glenn S. Smith, "Hybrid Wheat: What Is It?" *North Dakota Farm Research* 22 (March-April 1963): 13-15.
25. Robert L. Harrison, Mark Jendro, and J. F. Carter, "A Progress Report: The Agronomy Seed Farm," *North Dakota Farm Research 23* (November-December 1963): 8-9. See also David C. Ebeltoft, "The Seedstocks Project of the North Dakota State University, 1961-1981" (unpublished manuscript in the author's possession).
26. K. A. Gilles, "Office Memo," n.d., box 24, "Vice President for Agriculture K. A. Gilles (1)" file, AES Administrative Files, State Archives; interview with Richard Crockett, January 19, 1988; and Ebeltoft, "Seedstocks Project," 13 and 1.
27. Letter from Arlon G. Hazen to Kent Jones, June 13, 1968, box 165, "Misc. State Offices and Officials" file, p. 2, AES Administrative Files, State Archives.
28. Donald E. Anderson, "The Value of Quality Premiums Paid to North Dakota Wheat Farmers," *North Dakota Farm Research* 27 (May-June 1970): 6.

29. Letter from Arlon G. Hazen to K. A. Gilles, September 24, 1969, box 19, "Vice President, Agriculture" file, AES Administrative Files, State Archives.
30. Memorandum from J. F. Carter to Ten Members of the Agronomy Department, December 19, 1969, box 17, "Crop Quality Council" file; and letter from Mark Andrews to the Closing Committee of the Minneapolis Grain Exchange, January 13, 1970, box 19, "Misc. State Offices and Officials (incl. U.S. Congressman [*sic*] from N.D.)" file, AES Administrative Files, State Archives.
31. "1973 Wheat Program Not to Have Discount Program for Listed Undesirable Varieties," *U.S. Department of Agriculture News*, June 30, 1972.
32. Letter from K. A. Gilles to George Wilkins, March 2, 1970, box 19, "Vice President, Agriculture" file, p. 2; letter from Oben Gunderson, Jr., to Arthur Schultz, January 20, 1971, box 18, "Vice President, Agriculture" file, p. 1; and letter from Arden Burbidge to Laurel Loftsgard, May 8, 1972, box 20, "President Loftsgard-NDSU" File, p. 1, AES Administrative Files, State Archives.
33. Arlon G. Hazen, "From the Director," *North Dakota Farm Research* 28 (January-February 1971): 2; and letter from Arlon G. Hazen to L.D. Loftsgard, May 17, 1972, Box 20, "President Loftsgard-NDSU" File, p. 2, AES Administrative Files, State Archives.
34. "Office Memo, Confidential. Re: Meeting With Crop Quality Council and USDA Wheat Investigations Leader," March 4, 1971, box 18, "Vice President-Agriculture" file, p. 2; letter from Arlon G. Hazen to Clarence W. Sande, March 11, 1971; box 19, "AGSCO" file, p. 2, AES Administrative Files, State Archives; and J. F. Carter, "Perspectives in Hard Red Spring Wheat Breeding and Marketing," *North Dakota Farm Research* 29 (May-June 1972): 18.
35. Arlon G. Hazen, "From the Director," *North Dakota Farm Research* 30 (March-April 1973): 2 and 44; and Glenn S. Smith, "Changes in North Dakota Hard Red Spring Wheat Varieties, 1900-1977," *North Dakota Farm Research* 35 (May-June 1978): 17. For a discussion of Hazen's decision to release Olaf see Carter, "Agronomy."
36. Interview with Orville Banasik, July 23, 1987, and interview with Bert D'Appolonia, December 21, 1987.
37. Hugh Dufner and Gordon Erlandson, "The Growing Red River Valley Edible Bean Industry," *North Dakota Farm Research* 35 (November-December 1977): 10; and *North Dakota Historic Estimates: Selected Crops, 1955-1975* (Fargo: North Dakota Crop and Livestock Reporting Service, October 1980): 80.
38. J. J. Feight, "Bean Research: Against the Odds," *North Dakota Farm Research* 40 (November-December 1982): 8.
39. *North Dakota Historic Estimates . . . 1955-1975*: 39.
40. Arlon G. Hazen, "From the Director," *North Dakota Farm Research* 29 (July-August 1972): 2.

41. *North Dakota Livestock: County Estimates, 1925-1961* (Fargo: North Dakota Crop and Livestock Reporting Service, December 1961): 1, 20, 29, 44, and 50; and *North Dakota Crop and Livestock Statistics: Annual Summary for 1975* (Fargo: North Dakota Crop and Livestock Reporting Service, May 1976): 47.
42. See, for example, the letter from Arlon G. Hazen to Paul Emanuel, January 19, 1972, box 20, "Inquiries Answered" file, AES Administrative Files, State Archives; and interview with Verlin Johnson, January 9, 1986. For a full discussion of the relationship between the NDAES and dairying in North Dakota see my article "Promotion of Dairy Farming by the North Dakota Agricultural Experiment Station," *Agricultural History*, 63 (Spring 1989): 174-86
43. Interview with William Dinusson, July 23, 1987.
44. Dinusson interview; and letter from Arlon G. Hazen to Delbert Moore, July 25, 1974, box 23, "North Dakota Stockmen's Association" file, pp. 1-2, AES Administrative Files, State Archives.
45. Marvin J. Lemke and Laurel D. Loftsgard, "Land Prices Continue Up in 1960," *North Dakota Farm Research* 21 (March-April 1961): 12. See also Walter Wilson and Reuben Engelking, "Trends in North Dakota Agriculture," North Dakota Agricultural Experiment Station, *Bimonthly Bulletin* 19 (July-August 1957): 12-17.
46. Fred R. Taylor, "Changes in Number and Size of North Dakota Farms," *North Dakota Farm Research* 21 (September-October 1960): 18.
47. Thor A. Hertsgaard, "Distribution Patterns for Dressed Poultry," North Dakota Agricultural Experiment Station, *Bulletin 431* (Fargo: March 1961); Fred R. Taylor and David C. Nelson, "Factors Affecting the Transportation of North Dakota Grain," North Dakota Agricultural Experiment Station, *Bulletin 430* (Fargo: December 1960); and Laurel D. Loftsgard, "New Aid for Farm Planning," *North Dakota Farm Research* 21 (July-August 1959): 8-9.
48. Fred R. Taylor and Warren Tewksbury, "North Dakota's Potentials for a Straw Pulping Industry," North Dakota Agricultural Experiment Station, *Bulletin 406* (Fargo: April 1957); Ronald G. Fraase and Donald E. Anderson, "An Analysis of the Feasibility of Establishing Malt Plants in North Dakota," North Dakota Agricultural Experiment Station, *Bulletin 487* (Fargo: November 1960); Delmer L. Helgeson, David W. Cobia, Randal C. Coon, Wallace C. Hardie, LeRoy W. Schaffner, and Donald F. Scott, "The Economic Feasibility of Establishing Oil Sunflower Processing Plants in North Dakota," North Dakota Agricultural Experiment Station, *Bulletin 503* (April 1977); Rex W. Cox, Jerome E. Johnson, and Lowell K. Cook, "Survey of Resident and Non-Resident Tourist Groups," North Dakota Agricultural Experiment Station, *Bulletin 482* (Fargo: December 1969); and F. Larry Leistritz and Thor A. Hertsgaard, "Coal Development in North Dakota: Effects on Agriculture and Rural Communities," *North Dakota Farm Research* 31 (September-October 1973): 3-9.

49. Arlon G. Hazen, "From the Director," *North Dakota Farm Research* 32 (May-June 1975): 20; Hazen to Jones, June 13, 1968; and letter from Arlon G. Hazen to George R. Sollenberger, January 9, 1976, box 24, "Insurance Letters and Forms" file, p. 1, AES Administrative Files, State Archives.
50. Letter from Hazen to Posel, February 23, 1977; and letter from Arlon G.Hazen to Lloyd B. Omdahl, July 1, 1976, box 25, "Dept. of Accounts and Purchases" file, p. 2, AES Administrative Files, State Archives.
51. Letter from K. A. Gilles to Jack Dahl, Harold Hofstrand, and Joe Peltier, March 23, 1973; and letter from Milton R. Young to Arlon G. Hazen, February 22, 1973, box 21, "Fiscal 1974 Budget" file, AES Administrative Files, State Archives.
52. Hazen to Sollenberger, January 9, 1976, p. 1.
53. For a full discussion of the challenges to the system during the sixties and seventies see my essay, "Publicly Sponsored Agricultural Research in the United States from An Historical Perspective," in Kenneth A. Dahlberg, ed., *New Directions for Agriculture and Agricultural Research: Neglected Dimensions and Emerging Alternatives* (Totowa, N.J.: Rowman and Allanheld, 1986), 122-27.
54. Rachel Carson, *Silent Spring* (Boston: Houghton Mifflin, 1962).
55. Jim Hightower, *Hard Tomatoes, Hard Times: A Report of the Agribusiness Accountability Project on the Failure of America's Land Grant College Complex* (Cambridge, Mass.: Schenkman, 1973): xxvi.
56. These were remarks of the "Pound Committee," contained in the *Report of the Committee on Research Advisory to the U.S. Department of Agriculture* (Washington: National Academy of Sciences, 1972), 11 and 18.
57. Letter from Arlon G. Hazen to Clarence L. Jensen, March 14, 1975, box 23, "Inquiries Answered" file, p. 2, AES Administrative Files, State Archives. For general reaction within the system see André and Jean Meyer, "Agriculture, the Island Empire," *Daedalus* 103 (Summer 1974): 83-95; and Don F. Hadwiger, *The Politics of Agricultural Research* (Lincoln: University of Nebraska Press, 1982).
58. Letter from Arlon G. Hazen to Ralph K. Huitt, August 3, 1972, box 21, "NDSULGC-National Association of State universities and Land Grant Colleges" file, AES Administrative Files, State Archives. See also the letter from Arlon G. Hazen to K.A. Gilles, May 3, 1976, box 25, "Vice President for Agriculture K.A. Gilles" file, AES Administrative Files.
59. For the planning aspects of the Food and Agriculture Act of 1977 see Norwood Allen Kerr, "Institutionalizing the New Agenda: The State Agricultural Experiment Stations, 1977-1981," *Agricultural History* 62 (Spring 1988): 279-95.
60. Arlon G. Hazen, "Comments on the President's Budget Submitted to Congress for the Fiscal Period July 1, 1966-June 30, 1967," February 23, 1966, box 14, "Legislative Research Committee" file, p. 3, AES Administrative Files,

State Archives.

61. Letter from Milton R. Young to Arlon G. Hazen, March 14, 1968, box 16, "Misc. Offices and Officials" file; letter from Arlon G. Hazen to Mark Andrews, January 29, 1971, box 18, "Misc. State Offices and Officials (incl. U.S. Congressman [*sic*] from N.D.)" file, AES Administrative Files, State Archives; and Kerr, "Institutionalizing the New Agenda:" 281.
62. Letter from H. Roald Lund to William D. Ruckelshaus, February 4, 1971, Box 18, "Environmental Protection Agency" file; letter from John D. Nalewaja to Ted Schultz, July 8, 1976, box 25, "Inquiries Answered" file; letter from Arlon G. Hazen to Fred W. Clayton, July 19, 1976, box 25, "Inquiries Answered" file, p. 1, AES Administrative Files, State Archives; and interview with J.T. Schultz, September 29, 1987.
63. "Summary of Some Major Research Projects In the Agricultural Experiment Station," *North Dakota Farm Research* 31 (January-February 1974): 13-45; and "Summary of Some Major Research Projects in the Agricultural Experiment Station," *North Dakota Farm Research* 33 (January-February 1976): 5-43.
64. Letter from Wayne J. Colberg to Enoch Norum et. al., November 3, 1964, box 14, "Extension Service" file, AES Administrative Files; letter from Arlon G. Hazen to Mark Andrews, February 20, 1974, box 22, "U.S. Congressmen" file, AES Administrative Files, State Archives; K.A. Gilles, "The Environment: An Administrator's View," *North Dakota Farm Research* 28 (March-April 1971): 3; and letter from Edwin C. Becker to Arlon G. Hazen, May 3, 1971, box 18, "Miscellaneous Material" file, AES Administrative Files.
65. Callenbach interview.
66. Callenbach interview.
67. Interview with Charlotte Bennett, September 12, 1985.

CHAPTER 9

TOWARD THE FUTURE

1. Interview with Verlin Johnson, January 9, 1986. For Lund's background and philosophy see Don Lilleboe, "Dr. H. Roald Lund," *The Binford Guide* (October 1987): 13-17 and 38-39.
2. Interview with H. Roald Lund, December 22, 1987; and interview with Bryce Streibel, June 17, 1987.
3. Lund interview; and interview with Keith Bjerke, June 16, 1987.
4. H. R. Lund, "Director's Column," *North Dakota Farm Research* 36 (March-April 1979): 12.
5. Jerome E. Johnson, "Upward Trend in 1981 Farmland Values," *North Dakota*

Farm Research 39 (January-February 1982): 8.

6. Jerome E. Johnson, "Farmland Values Declined More in 1984," *North Dakota Farm Research* 42 (January-February 1985): 7; and idem, "North Dakota Farmland Values in 1985," *North Dakota Farm Research* 43 (January-February 1986): 6.
7. Interview with William Guy, July 2, 1987.
8. Interview with Arthur Link, June 30, 1987; interview with James Brokaw, June 18, 1987; interview with Scott Stofferahn, June 19, 1987; and interview with Charles Edgerly, July 27, 1987.
9. *North Dakota Farm Research* 42 (July-August 1984). For the response of agricultural experiment stations, see Mark Friedberger, *Farm Families and Change in 20th-Century America* (Lexington: University Press of Kentucky, 1988). See also F. Larry Leistritz, Brenda L. Ekstrom, Harvey G. Vreugdenhil, and Arlen G. Leholm, "Off-Farm Income of North Dakota Farm Families," *North Dakota Farm Research* 43 (January-February 1986): 39-45; and Harriet K. Light, Richard Hanson, and Doris Hertsgaard, "The Work of North Dakota Farm Women," *North Dakota Farm Research* 41 (September-October 1983): 25-26.
10. John F. Mittleider, Randal C. Coon, Thor A. Hertsgaard, Delmar L. Helgeson, and Eddie V. Dunn, "Feasibility Studies Have High Payoff in North Dakota," *North Dakota Farm Research* 41 (September-October 1983): 40-43; Kenton R. Kaufman and Harold J. Klosterman, "A Highway Test of Gasohol," *North Dakota Farm Research* 37 (July-August 1979): 18-22; John Walter, Paul Aakre, and John Dury, "The North Dakota 'Flower Power' Project," *North Dakota Farm Research* 39 (May-June 1982): 3-6; and Lund interview.
11. Interview with Myron Just, June 19, 1987.
12. Interview with Thomas J. Conlon, July 1, 1987.
13. Interview with Lois Jones, June 16, 1987.
14. Interview with Frank Kubik, July 1, 1987; Streibel interview.
15. Interview with George Sinner, July 2, 1987.
16. Interview with John Nalewaja, December 21, 1987; Bjerke interview.
17. Interview with Gary Secor, December 21, 1987.
18. Interview with Richard Crockett, January 19, 1988; interview with Clark Jenkins, July 21, 1987; and interview with Howard Olson, June 21, 1987.
19. Interview with Robert Johansen, August 17, 1987; interview with George Pratt, August 21, 1987; interview with Jack Carter, October 1, 1987; and Conlon interview.
20. Interview with Richard Kiesling, August 17, 1987.
21. Interview with Ross Wilkinson, December 23, 1987.
22. Interview with Richard Golberg, December 22, 1987.
23. Secor interview.
24. Streibel interview; interview with J.L. Ozbun, July 26, 1989.
25. Guy interview.

26. Streibel interview.
27. Pratt interview.
28. Conlon interview; Olson interview; and Kubik interview.
29. Kiesling interview.
30. Crockett interview; and Ozbun interview.
31. Secor interview; Johansen interview; interview with Orville Banasik, July 23, 1987; and Lund interview. For a broader perspective on this issue see Frederick H. Buttel, "Agricultural Research and Farm Structural Change: Bovine Growth Hormone and Beyond," *Agriculture and Human Values* 3 (Fall 1986):88-98.
32. Interview with Christian Jensen, July 11, 1984; and Johansen interview.
33. Jensen interview.
34. H. E. Stockbridge, *First Annual Report of the North Dakota Agricultural Experiment Station to the Governor* (Fargo: Daily Republican Printing, 1891): 11; and Lund interview.

INDEX

The following references were inadvertently omitted from the index: